International Conference for Free and Open Source Software for Geospatial (FOSS4G 2013)

OSGeo Journal Volume 13, Issue 1

Nottingham, United Kingdom
17-21 September 2013

Editors:

Barend Köbben

ISBN: 978-1-7138-1899-1

Printed from e-media with permission by:

Curran Associates, Inc.
57 Morehouse Lane
Red Hook, NY 12571

Some format issues inherent in the e-media version may also appear in this print version.

OSGeo
9450 SW Gemini Dr. #42523
Beaverton, Oregon
United States
97008

www.osgeo.org

Additional copies of this publication are available from:

Curran Associates, Inc.
57 Morehouse Lane
Red Hook, NY 12571 USA
Phone: 845-758-0400
Fax: 845-758-2633
Email: curran@proceedings.com
Web: www.proceedings.com

Free and Open Source Software for Geospatial (FOSS4G) Conference Proceedings

Volume 13 *Nottingham, UK* Article 14

2013

Volume Introduction Letter

Barend Köbben
ITC–University of Twente

Recommended Citation

Köbben, Barend (2013) "Volume Introduction Letter," *Free and Open Source Software for Geospatial (FOSS4G) Conference Proceedings*: Vol. 13 , Article 14.
DOI: https://doi.org/10.7275/R5FX77MH
Available at: https://scholarworks.umass.edu/foss4g/vol13/iss1/14

FOSS4G

NOTTINGHAM 2013

OSGEO Journal Vol. 13 – Feb 2014

Geo For All

From the Editor...

by Barend Köbben

Welcome to this issue of the OSGeo Journal, comprising twelve research papers selected from the submissions to the Academic Track of FOSS4G 2013, the global conference for Open Source Geospatial Software, which took place in Nottingham (UK), from 17 to 21 September 2013.

FOSS4G is not an academic conference. The core audience has always been the people who make up the *open source communities*: The people that develop, create and craft the open source geo-spatial software. The actual applications are the glue which binds the community together; the aim of the FOSS4G community is to enable and enfranchise anyone to harness the power of geo-spatial software, regardless of their economic status. To acknowledge this, and to not create an isolated, exclusive, part of the conference, we scheduled presentations of the papers clustered with other, non-academic, papers based on subject matter. By this, we hope to have generated attention for academic input in the community and to cross-pollenate with industry, developers and users.

This year, the Academic Track motto was "Science for Open Source, Open Source for Science", and we made an effort to gather papers describing both the use of open source geospatial software and data, in and for scientific research, as well as academic endeavours to conceptualize, create, assess, and teach open source geospatial software and data. We hope that the papers in this issue reflect that idea.

We thank the reviewers (listed on the imprint page at the end of this issue) for making the Academic Track possible.

Barend Köbben, ITC–University of Twente

FOSS4G2013 Proceedings Editor OSGeo Journal
`http://osgeo.org`
`b.j.kobben@utwente.nl`

Contents of this volume:

A New Zealand case study

Open Source, Open Standards, Open Data

by Brent Alexander Wood

NIWA, New Zealand. pcreso@pcreso.com

Abstract

The National Institute for Water and Atmospheric Research (NIWA) is New Zealand's leading agency providing freshwater, ocean, climate, atmosphere and fisheries related research. Open Source software is widely used internally, both infrastructurally and in desktop systems.

In 2011, the New Zealand Government passed "The Declaration on Open and Transparent Government". This requires central government agencies to make taxpayer funded information freely available to the public, and encourages regional and local government, as well as agencies such as NIWA to comply. NIWA works closely with central and regional government, utilities, NGO's and primary industry, making information discovery and delivery using common and open standards critical.

NIWA is using Open Source applications to meet these open data discovery and delivery requirements. Open Geospatial Consortium (OGC) standards compliance ensures interoperability. Standards adopted to date include SFS (Postgis), CSW (Geonetwork), WMS/WFS (Mapserver, Geoserver, Openlayers, Quantum GIS, Quantum Map) and SOS (52°N, Quantum Map). Some proprietary applications are also used. These are also OGC compliant and fit within NIWA's OGC based architecture. This paper describes the role that open source software and open standards play in NIWA's strategies and architecture for environmental information management, discovery and delivery and gives implementation examples.

Keywords: environmental data, interoperability, NIWA, OGC, open data, open source, OSGEO, web services.

Introduction

This paper presents several examples of using Open Source GIS related tools to manage and deliver environmental data captured by a variety of research programmes undertaken by the National Institute for Water and Atmospheric Research in New Zealand (NIWA, http://www.niwa.co.nz).

The systems are fully compliant with open industry standards developed by the Open Geospatial Consortium (OGC) which are the basis of New Zealand government standards for spatial data interoperability (NZGO, 2011), to comply with the recommendations of the New Zealand Government's 2011 Declaration on Open and Transparent Government:

"To support this declaration, the government asserts that the data and information it holds on behalf of the public must be open, trusted and authoritative, well managed, readily available, without charge where possible, and reusable, both legally and technically. Personal and classified data and information must be protected.

Public service and non-public service departments are directed and State Services agencies encouraged to commit to the release of high value public data for re-use in accordance with the declaration and principles."

NIWA has therefore a responsibility to ensure the publicly funded environmental data it holds, as well as the results of research and analyses on these data, are made readily available. This is being done primarily using Open Source tools to implement data (and metadata) discovery and delivery capabilities for NIWA held data.

NIWA's environmental information needs.

NIWA needs integrated and interoperable systems to manage its various and heterogeneous environmental data and information holdings, and to provide search and access facilities to enable discovery and delivery, both internally, within its various science centres, and externally, for clients, collaborators, other agencies and the public.

NIWA's internal requirements

NIWA is New Zealand's leading environmental science provider in the aquatic, climate, atmospheric and fisheries domains. As a New Zealand Crown Research Institute (CRI), NIWA operates as a commercial business, although the New Zealand government is the only shareholder. Internally, NIWA is divided into science centres, each focused on one of NIWA's environmental science domains. Given the highly inter-related nature of physical, chemical and biological data in fresh water, coastal and oceans science, and of climate research on freshwater availability, etc, there is within NIWA a critical need for information management, discovery and delivery systems that support all the areas of science NIWA undertakes, and are not domain specific silos within the organisation.

Virtually all environmental data has a critical spatial component, so NIWA's environmental data solutions are very much spatial data soutions, hence the significant use of OSGEO applications and tools in NIWA (as well as other open source tools), for integrated spatial data management, discovery and delivery systems.

Client interoperability requirements

NIWA clients include a wide range of goverrnment agencies (central, regional and local), utilities (eg. wind and hydro power generation, electricity transmission), industry (e.g. agriculture, irrigation). NIWA also collaborates with international agencies & has international obligations on behalf of New Zealand. NIWA's data management and delivery systems therefore need to support a wide range of relationships with local, national and international agencies.

To enable data and information discovery and delivery services accross such a wide range of users and clients from internal, regional, national and international levels, a high degree of compliance with industry standards across several domains is necessary.

NIWA is following the OGC and ISO standards recommended for New Zealand Government agencies, and is both internally and collaboratively developing a range of standards, tools and systems for users and clients which provide suitably interoperable information discovery and delivery facilities. These make use of Open Source tools and components from the Open Source Geospatial Foundation (OSGEO), but include other Open Source applications and tools, as well as some commercial applications.

Where there are no New Zealand recommended standards for particular domains, NIWA is collaborating with other New Zealand agencies to adopt or devise suitable standards, and is also developing working applications as exemplars for other New Zealand agencies.

NIWA's history with open source

NIWA has a long history of supporting and internal use of open source applications. In support of their research projects, various NIWA staff have used open source applications for some years. Also, Linux servers have been used as web and database servers for some time, generally being used to replace older UNIX servers during the 1990's. In the geospatial arena, given New Zealand's location in the world, NIWA has to contend with well known issues relating to the 180° meridian. To help work around these, NIWA funded the original implementation of the ST_Shift_Longitude() Postgis function (`http://postgis.net/docs/manual-2.0/ST_Shift_Longitude.html`), as well as the cs2cs parameter "+lon_wrap" for Proj.4 (`http://trac.osgeo.org/proj/wiki/GenParms#lon_wrapover-LongitudeWrapping`). NIWA also developed the new vector format for Generic Mapping Tools (GMT) v5.0 (Wessel et al, 2011), along with funding the GDAL/OGR driver for this format, enabling GMT to interoperate with many other open source GIS related applications and tools. NIWA also supports other open source tools such as GERRIS (Popinet, 2003), CYLC (Oliver, 2012) and has twice been a finalist in the New Zealand Open Source Awards.

Data management and delivery strategy

To support information needs as outlined above NIWA has adopted a strategy for information management, discovery and delivery. Part of the strategy is the support of core infrastructure components through standardized database solutions; these can be open source or commercial, depending on the science and user needs. Metadata and Data are exposed using (open or industry) standard compliant web services where possible. NIWA's strategy for meeting its spatial data interoperability needs is largely based on OGC standards enabled by open source applications. This approach ensures the delivery systems are decoupled from the underlying data management systems (see Figure 1).

The main information infrastructure components used at NIWA:

- Metadata Catalog (Geonetwork)
- Station Catalog (Postgis)
- Image Catalog (AtlasMD - commercial product)

- Spatial Layer Archive (ESRI – commercial product)
- Taxa Catalog and Information System (Posgis)
- Timeseries databases (various bespoke, commercial and open source products)
- Observation databases (Postgis)
- Marine and atmospheric data archives (netCDF, Postgis)
- Product Catalog (commercial and open source products)
- Webservice solutions (Mapserver, Geoserver and industry standard solutions)
- Publishing and Reporting System (Jasper)
- Web delivery (Openlayers and others)

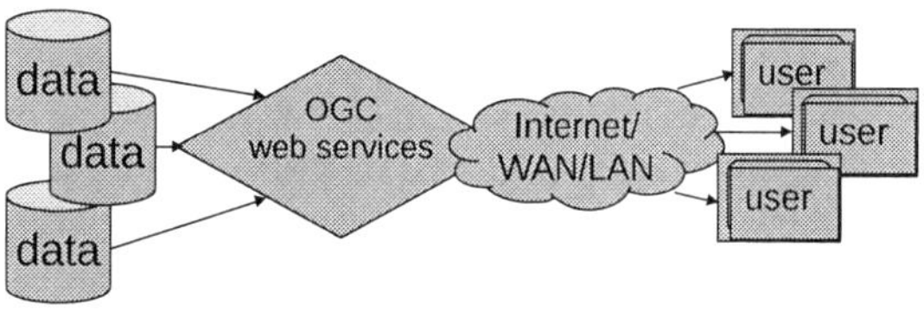

Figure 1: Internally managed data (or metadata) from multiple data stores are provided using OGC compliant (CSW, WFS, WMS, SOS) and other web services to the internet, enabling internal and external consumption, via both web and desktop clients.

As one can see NIWA's information infrastructure is based on a complex system of interacting components, with open source and open standards solutions making up a large component of that infrastructure.

Use of open source and open standards for NIWA's environmental information discovery and delivery

In 2011, NIWA established an Environmental Information Centre (EIC, https://www.niwa.co.nz/our-science/ei). This was a significant change for NIWA, as all previous NIWA Research Centres were focused on specific science domains. The new centre has a mandate to work across the science domains, fulfilling a strategic role and supporting data monitoring, management and delivery capabilities throughout NIWA, as well as working with clients and other stakeholders in New Zealand's environmental research and management arena to facilitate data sharing and re-use.

The EIC is supporting the use of open source tools to achieve its goals, and in line with the freedoms that open source provides, supporting the re-use of the systems and standards both inside NIWA and externally to clients and other agencies.

The following sections describe examples for NIWA's various uses of open source applications and open standards to enable this strategy.

IPY/CAML metadata catalogue

One of the first open source applications formally used by NIWA to meet contracted requirements for data discovery and delivery (rather than as an in-house research tool) was Geonetwork (`http://geonetwork-opensource.org/`).

NIWA undertook a research voyage to the Ross Sea as part of the International Polar Year/Census of Antarctic Marine Life (IPY/CAML) during 2008, which included a requirement to make all datasets and reports publicly available.

At this time, Land Information New Zealand (LINZ), the government agency responsible for land titles, geodetic and cadastral survey systems, topographic information, hydrographic information, managing Crown property and a variety of other functions, was also establishing a metadata catalogue, a national catalogue for geospatial data holdings (`http://www.geodata.govt.nz`), in conjunction with the government agency responsible for science funding, the Ministry of Research, Science and Technology (MORST). This was also based on an underlying Geonetwork catalogue, with website development undertaken by the local web development company, Silverstripe, using their open source CMS of the same name (`http://www.silverstripe.org/`). NIWA chose to also contract the initial implementation and hosting of its catalogue to Silverstripe.

A Geonetwork metadata catalogue was established and populated to successfully provide discovery and delivery capability. This has since been moved to sit within a more generic web portal established to provide information discovery and delivery facilities for NIWA research projects, (`http://www.os2020.org.nz/new-zealand-ipy-caml-project/`).

One side effect of these initiatives, in line with the increasing New Zealand Government's support of open data, was a workshop on metadata catalogues run by LINZ, using Geonetwork as the enabling tool. This was attended by representatives from central and regional government, universities, NIWA and other CRI's, utilities and private businesses, and was a seminal influence in the ongoing development of information discovery and delivery systems in New Zealand.

Ocean Survey 20/20 Bay of Islands web portal.

In 2010, NIWA undertook an extensive multidisciplinary survey off the north east coast of the North Island (see Figure 2) with a particular focus on the Bay of Islands, as part of the New Zealand Ocean Survey 20/20 initiative.

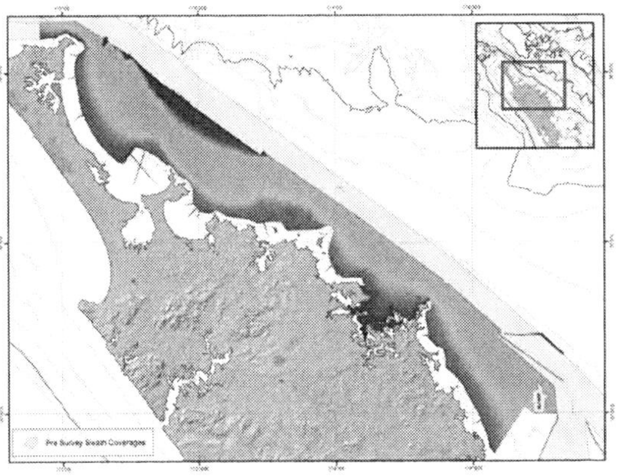

Figure 2: Map showing the Bay of Islands survey area, including the seabed bathymetry from shallow inshore areas to water up to 50 metres deep.

This project included a requirement to provide public access to the project's data and reports, as well as describing and documenting the survey and its many participating agencies and staff. A web portal was chosen as the appropriate solution, and was developed using the following open source tools (mostly from the OSGEO stack) enabling both server and client capability:

- Geonetwork metadata catalogue for information discovery and download
- Postgis spatial database server
- Mapserver to provide OGC WMS/WFS contextual and data layers
- Openlayers as the web map client
- Silverstripe as the enabling CMS, with both:
  1. a map administration tool supporting an embedded Openlayers WMS/WFS client
  2. a CSW client providing embedded user friendly access to the Geonetwork metadata catalogue.

This solution is based on a client/server architecture which allows the data management and delivery (server) functionality to be separated from the portal (client). Using OGC web services enables the portal to display content from various independent data repositories, including Postgis databases and ESRI ArcServer/SDE data stores. This also enables other agencies and users to reuse the WMS and WFS layers in their own applications, web sites or desktop mapping applications as desired.

Silverstripe developed an Openlayers administration module for the open source Silverstripe CMS for this project. This tool allows the web site administrators to manage the web portal's map map pages, by creating and managing maps and map layers. Layers are assigned to maps, and are defined as a URL for an OGC WMS or WFS service, along with information required to manage the map and the layers it comprises. This tool separates the roles of GIS and web site administration. GIS staff are responsible for systems managing the spatial data and serving the map layers, and web site administrators manage the maps and layers within the web site. The tool also enables re-labeling of WFS data fields for feature queries, and integrates metadata catalogue searches from the map, by allowing a set of keywords to be associated with a map layer, with the ability to for a user to request a catalogue search from the map interface, allowing direct access to relevant reports, datasets, map images and other documents directly from the map interface.

Some of the datasets and samples captured during the survey were analysed and made available as OGC web service map layers using ESRI ArcServer. Some 20,000 seabed photographs captured during the survey are managed in Atlas, a commercial digital asset management tool used by NIWA as an image database. NIWA, Silverstripe and the Atlas developers implemented a custom web API to provide the Atlas images to the portal map via OpenLayers (see Figure 3).

The Silverstripe CMS also manages web pages describing the methods used during the survey, including physical gear which was deployed, manual sampling methods, and for analytical results, descriptions of the analyses undertaken. The embedded Openlayers administration tool includes a link to these pages, so the user can view the layer's associated method information directly from the Openlayers layer menu, simply by clicking on the layer name (see Figure 4). This provides a map-centric approach to the web portal, with users able to access a variety of non-map content directly from the map.

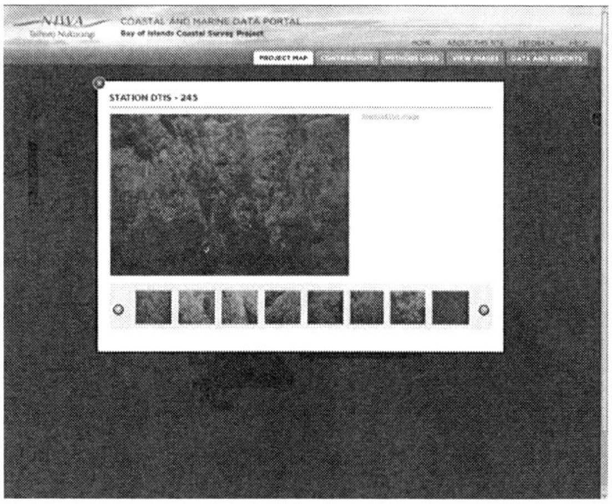

Figure 3: A custom web service allows users to view over 20000 seabed images from the Bay of Islands survey by clicking on a point displayed inthe Openlayers map.

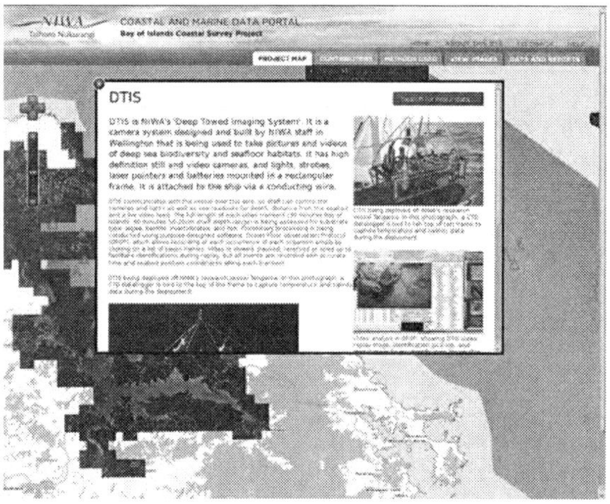

Figure 4: Clicking on the name of a layer in the OpenLayers layer menu displays the layer method page from the CMS, as configured in the Silverstripe OpenLayers administration module, providing a map-centric user interface embedded within the CMS.

The original Silverstripe portal has since been redesigned to support multiple projects within a single implementation, as well as being able to be re-deployed as a completely new portal instance where appropriate. This maximises the potential for re-use of the portal as a complete spatial data discovery and delivery system, either by adding new projects to an existing portal, or by redeploying new instances of the portal. It is now being used as a generic web interface for NIWA research projects which require a web presence or interface (`http://www.os2020.org.nz/`).

OS2020 Chatham/Challenger survey web portal

Figure 5: The "species picker" allows the user to select a species, to display two separate layers, the observed locations of the species as a WFS layer, on the Chatham Rise to the east of New Zealand and the Challenger Plateau to the west, and the modelled, predicted species distribution, via WMS.

Another survey within the OS20/20 research programme examined the biodiversity of the Chatham Rise and Challenger Plateau, off the east and west coasts of New Zealand respectively. After the survey was completed, an additional contract required NIWA to make the data and reports available to the public by reusing the successful Bay of Islands portal. Additional biodiversity content required some enhancements to the Silverstripe map administration tool, noteably regarding a "species picker" facility (see Figure 5) for users to select from a list of available species, providing the ability to display both a map layer showing where the species was observed, as well as (where available) a layer showing the predicted distribution. Following the client/server via web service architecture of the portal, the list of species is also provided via a web service, so is driven from the server databases. This project also required the seamless display of data across the 180° meridian, enabled using Postgis, Mapserver and Openlayers.

The New Zealand Marine Biosecurity Porthole

In 2011, NIWA was approached by Biosecurity New Zealand to create a new instance of the portal to provide access to New Zealand marine biosecurity data. This was done (`http://www.marinebiosecurity.org.nz/`), and provides an example of the easy re-deployment and reuse of systems built using open source tools (see Figure 6). Biosecurity survey data was loaded into a Postgis database with Mapserver used to deliver the data via OGC WMS and WFS services. The system uses the Silverstripe CMS, along with both the Openlayers web mapping tool and the Geonetwork metadata catalogue to provide marine biosecurity data and reports to stakeholders in New Zealand.

Figure 6: The New Zealand Marine Biosecurity Porthole provides access to New Zealand's marine biosecurity data and related information for stakeholders.

NIWA Data Catalogue

NIWA's Environmental Information Centre has initiated several new projects and has continued to maintain and enhance existing ones, relying heavily on open source components, building on NIWA's experience with the Bay of Islands portal.

The NIWA Data Catalogue (`http://dc.niwa.co.nz/niwa_dc/srv/en/main.home`) was implemented to provide a searchable discovery facility for NIWA staff (internally) as well as clients and other users (externally). Based on our successful deployment of Geonetwork for project based information discovery and delivery systems, NIWA chose Geonetwork for its institutional catalogue.

NIWA has developed a custom tool to facilitate the entry of metadata records, called Postcard (see Figure 7). This presents the metadata fields to the user within a series of tabs, each covering a different aspect of the metadata, including "who", "where", "when", etc, which provides a user friendly metadata entry tool. The tool is able to automatically fill some fields based on the user id, and the knowledge that NIWA is the responsible agency. All records are saved to GeoNetwork, but are not published until approved by the NIWA metadata curator, who is notified by an email from the Postcard tool about each new record.

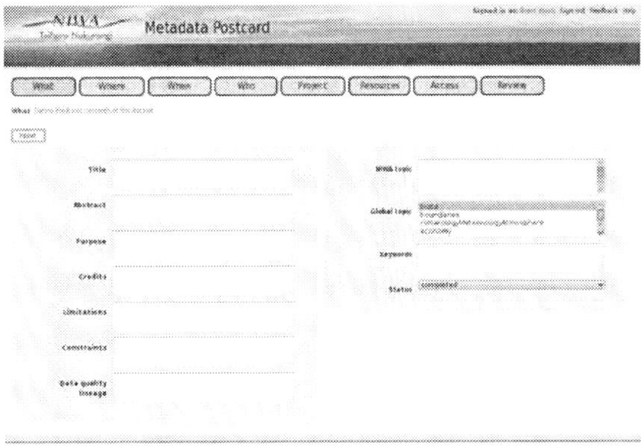

Figure 7: The NIWA postcard tool provides a user friendly interface for metadata entry. Tabs break up the content into subject areas, pick lists provide a simple controlled vocabularies for appropriate fields, The tool automatically emails NIWA's metadata curator of each record entered. The curator then checks the new record meets NIWA guidelines before enabling public access to it.

Records from NIWA's catalogue are automatically and selectively harvested into other catalogues, including national and international catalogues. NIWA has been pleasantly surprised with how often such catalogues are also using Geonetwork, making harvesting and interoperability relatively straightforward.

NIWA has adopted the Australian Marine Communities Metadata Profile as the preferred profile for its metadata. This is an ISO19115/19139 compliant profile, based on the ANZLIC (Australia and New

Zealand Land Information Council) profile, with enhancements for marine metadata (AODC, 2008), which has been supported by GeoNetwork almost since its inception.

NIWA Environmental Information browser

The Environmental Information Browser (EIB) is a web based facility (`http://ei.niwa.co.nz/`) enabling concurrent searches on several NIWA catalogues and databases which deliver information through OGC web services (see Figure 8). Users can direct searches against specific systems, or across several systems or datasets at once. Searches support keywords, as well as geographic and temporal constraints. The EIB is in a state of ongoing development as further databases and new functionality are added. The EIB itself is a bespoke application implemented using the open source Symfony framework, which makes extensive use of OGC web services to search and deliver content from the various underlying systems. The spatial databases it accesses generally implemented using Postgis, served as OGC WMS and WFS services via Geoserver or MapServer. The data catalogues are GeoNetwork instances searched using OGC CSW.

Given that the underlying NIWA systems accessed by EIB are providing CSW, WMS and WFS web services, other applications and agencies are able to seamlessly embed content from these NIWA systems in their own applications.

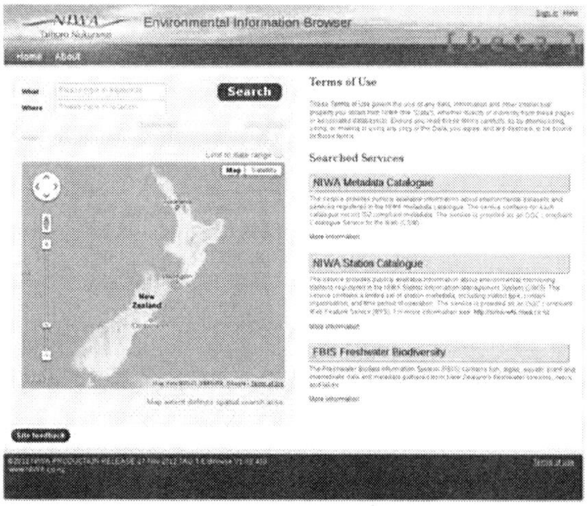

Figure 8: NIWA's Environmental Information Browser provides a single search facility across several systems at once, with spatial, temporal and keyword filters.

NIWA Taxonomic Information System

NIWA is developing three linked applications to provide a suite of taxonomic data capabilities for internal and external users, which together comprise the Taxonomic Information System (TIS, `https://tad.niwa.co.nz/`).

While taxonomic data are not inherently spatial themselves, they are critical information supporting a wide range of environmental science, which is decidedly spatial in nature. NIWA in integrating these with our spatial biodata systems, such as the web portals described above, so the taxonomic systems being developed and their open source components are included here.

Open source components used for these applications include:

- PHP The main programming language for the web interface (`http://php.net/`)
- The Symfony framework (`http://www.symfony-project.org/`)
- JQuery – javascript library for web user interfaces (`http://jquery.org/`)
- Jasper Reports (`http://community.jaspersoft.com/wiki/community-wiki`)
- D3 Data Driven Documents (`http://d3js.org`)
- PostgreSQL database (`http://www.postgresql.org`)
- Apache Webserver (`http://www.apache.org`)
- Apache Tomcat Java Servlet and JSP framework (`http://tomcat.apache.org`)
- Jenkins extendable open source continuous integration server (`http://jenkins-ci.org`)
- Apache Ant build scripting language (`http://ant.apache.org`)
- Suse Linux Enterprise Server (SLES 11) Linux Platform (`https://www.suse.com/`
- Apache Subversion source code revision control (`http://subversion.apache.org`)
- Firefox browser & Firebug (`http://www.mozilla.org`)

Taxonomic Reference System

Quality taxonomic information is critical to support NIWA's biodiversity, biosecurity and fisheries research, as well as the National Invertebrate Collection (NIC), New Zealand's primary museum collection of aquatic invertebrates. The Taxonomic Reference System (TRS) provides NIWA's link to the New Zealand Organism Register (NZOR), the national taxonomic reference database. TRS stores a subset of the NZOR content, those taxa relevant to NIWA, as an institutional reference dataset for staff. For taxa where NIWA is the national authority, TRS will provide authoritative taxa data to NZOR. TRS is implemented on a Postgres database.

Given the maritime and international nature of much of NIWA's biodata, the World Register of Marine Species (WoRMS, `http://www.marinespecies.org/about.php`) is also a valuable reference for staff, and TRS provides direct access to the WoRMS taxonomic hierarchy as well as NZOR, and the local NIWA version. (see Figure 9).

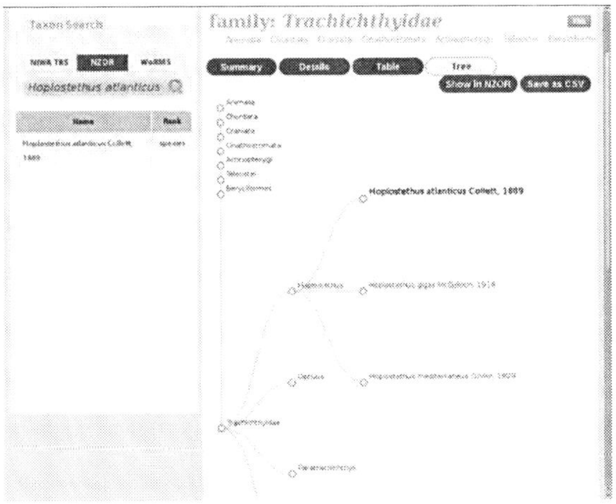

Figure 9: The NIWA Taxonomic Reference System (TRS) provides direct access to taxonomic information from NIWA's internal taxonomic database, as well as the New Zealand Organisms Register (NZOR) and the World Register of Marine Species (WoRMS).

Taxonomic Attribute Database

TAD, the Taxonomic Attribute Database, is a Postgres database storing taxon traits (attributes) which are associated with a TRS taxon via the TRS taxon ID. These traits include text, photographs, dates, etc. and often comprise the contents of various fields used for producing taxon based fact sheets or ID guides (see Figure 10). As with the Bay of Islands web portal, images are stored in the Atlas digital asset management application, and retrieved using a web service, the TAD record stores the Atlas image key.

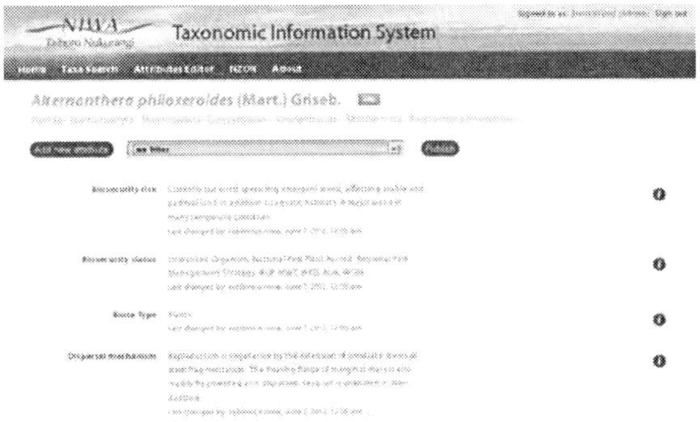

Figure 10: The Taxonomic Attribute Database (TAD) stores taxon trait information (such as text describing the biosecurity risk) for taxa recorded in TRS, and has a web based interface to manage and edit the content, which is accessible to authorised users both within NIWA and externally.

Publishing and Reporting System

The Publishing and Reporting System (PRS) manages templates describing the layout of such documents. It uses the open source version of Jasper Reports to create PDF or HTML documents from content stored in TRS or TAD, arranged according to the selected template in PRS. Several documents (templates) are described in PRS, with data in TRS/-TAD, including a New Zealand freshwater pests guide, a national freshwater fish atlas and an otolith guide (used for prey identification in fish, bird and marine mammal stomach contents, and to identify fish food species from middens in archeological research). Users can select a single species or list of species to include in the specified report (see Figure 11).

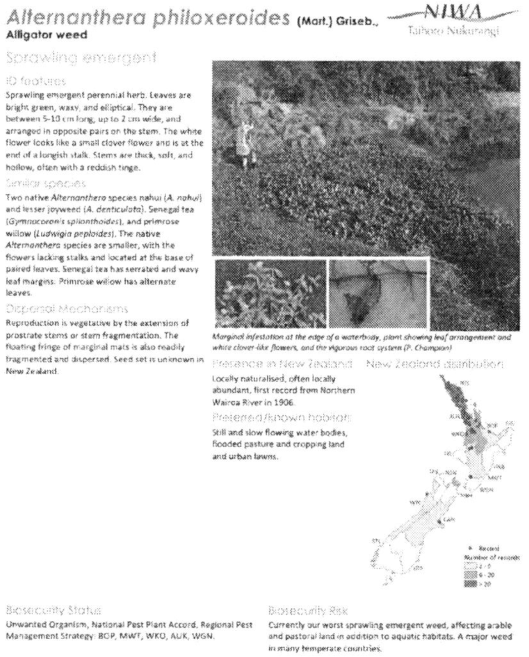

Figure 11: The NIWA Publishing and Reporting System generates documents such as this fact sheet for a freshwater pest species, with content retrieved from TRS and TAD, and the layout determined by a template.

NIWA Environmental Observation and Monitoring System

The NIWA Environmental Observation and Monitoring System (NEMO) is an attribute/value database implemented using Postgres. The generic database structure supporting NEMO is capable of storing and managing virtually any measurement or observation data, and several datasets that used to be managed independently have now been migrated to NEMO.

As NEMO has a variety of OGC web services already implemented on the underlying database, these services are available ready for use on any dataset added to NEMO.

To facilitate the loading of datasets into NEMO, NIWA has developed the NIWA Observations Ingestion System (NOIS) – a flexible framework for ingesting various types of observations from data providers, through web services.

NIWA Quantum Map

NIWA Quantum Map (see Figure 12) is a customised version of Quantum GIS provided by NIWA (`http://www.niwa.co.nz/software/quantum-map`), developed under contract by Sourcepole AG in Switzerland. Quantum Map has some custom functionality added to simplify access to data provided by OGC web services, and some of the more powerful Quantum GIS analytical capabilities have been removed, to provide a simpler mapping tool for users to access data provided by OGC web services by NIWA and other agencies.

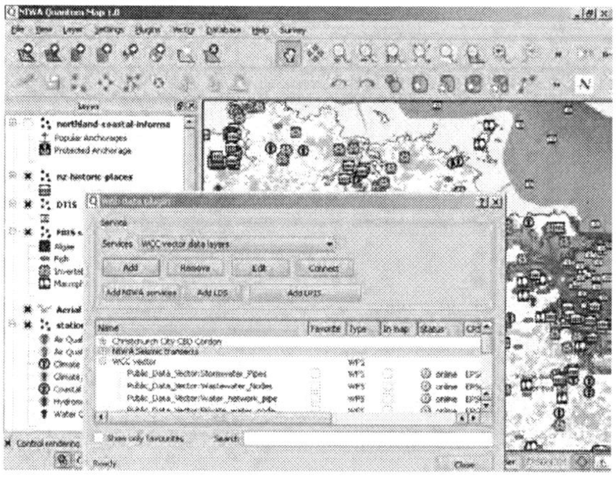

Figure 12: NIWA Quantum Map, showing a project with OGC WMS and WFS layers from NIWA's Ocean Survey 20/20 portal, as well as from other datasets and OGC services. The NIWA web data tool is shown in the foreground, listing WFS layers available from the Wellington City Council (WCC). These layers can be added to a user's "favourites" list, added to the map, or downloaded and cached locally.

The main NIWA tool provided in Quantum Map is an enhanced OGC client. This is able to retrieve a list of WMS and WFS service providers on the NIWA web site (`http://www.niwa.co.nz/ei/feeds/report`) and to open the map layers from these sources. It is also able to cache these layers (both WMS and WFS) locally for offline use.

Given NIWA's decision to use OGC web services for environmental data delivery and discovery, Quantum Map is an important part of the NIWA strategy for open data access. Having such a tool freely available means that by simply providing an OGC WMS or WFS service on a dataset, NIWA is enabling access for anyone using this client application (or any other OGC client). Developing web portals for every dataset is no longer necessary for users to discover, access and visualise NIWA spatial data. As can be seen at the above URL, not just NIWA datasets are listed and available for use with Quantum Map. Central and local government agencies, private businesses and other research agencies also provide map layers via OGC web services, and are readily available for Quantum map users.

An important subset of NIWA's environmental data includes time series data, from both climate and hydrometric stations, managed in NIWA's national climate and hydrometric databases respectively. During 2013 NIWA is developing OGC SOS (Sensor and Observation Service) web services for these databases, and will be providing SOS clients for these in Quantum Map. NIWA is discussing the best approach for delivering such time series data with stakeholders and interested parties, with the likely approach being a combination of WFS for discovery and SOS for delivery, much as described at FOSS4G2010 (Hollmann et al. 2010).

NIWA is also looking at the CSW support available for QGIS, and adapting this to enable Quantum Map to directly access NIWA's institutional Geonetwork catalogue, and the OGC web services described there.

NIWA hopes that improved plugin support provided in future Quantum GIS will enable the various NIWA tools from Quantum Map can also be made available as QGIS plugins, so that Quantum Map becomes simply a version of QGIS with preloaded plugins, supporting all the platforms that QGIS does.

Discussion

Much as described in the US Department of Defence's guidelines on the use of Open Source software, adapted below from (Scott et al, 2011), NIWA has found several advantages in using FOSS software for open standards compliant systems enabling data and metadata interoperability with clients, collaborators and stakeholders, including:

- Increased Agility/Flexibility: NIWA has full access to source code, in-house developers can work with the developer community to adapt existing tools to better meet NIWA needs, generally within the normal development process with no forking or bespoke systems required. Open source applications can be adapted to make use of common infrastructural facilities in NIWA, and therefore a more efficient implementation is achieved, much like an optimised institutional software suite. Support is available directly from the core development team, enabling rapid response times for bug fixes and enhancements.

- Less risk, better planning. Open source software projects generally maintain a public web site listing bugs, problems, requested enhancements, etc, as well as the fixes. Perusing such sites provides a wealth of detail regarding any problems, as well as the response times to solve issues etc. This sort of information is often impossible to find for commercial software.

- Faster delivery: Because developers only need to focus on changes to, and integration of, existing software capabilities they can significantly reduce the time to delivery for new capabilities.

- Increased Innovation: With access to source code for existing capabilities, developers and contractors can focus on innovation and the new requirements that are not yet met by the existingsource code capabilities, working within a community of existing developers and sharing both information.

Conclusion

NIWA has had a long and successful involvement with open source software, Recent New Zealand government Open Data initiatives support NIWA's goal to make data more accessible to clients, stakeholders and the public and other interested parties. NIWA has made a significant commitment to delivering open data, via open standards, implemented using robust open source applications.

NIWA's experiences using open source applications as research tools, and to provide environmental information management, discovery and delivery capability for institutional, national and international systems have been overwhelmingly positive. The wide support that exists for OGC standards in both the open source and proprietary GIS spaces has meant that a high level of interoperability exists for users of tools from both camps, enabling common frameworks despite different underlying tools.

Acknowledgement NIWA acknowledges the support and funding provided by Land Information New Zealand through the Oceans Survey 20/20 programme and the New Zealand Geospatial Office; the

Ministry for Primary Industries, the Ministry of Business, Innovation and Employment and the Terrestrial and Freshwater Biodiversity Information System (TFBIS) Programme managed by the Department of Conservation.

NIWA recognises and appreciates the contribution OGC has made to providing standards that enable environmental data interoperability, discovery and delivery, and the contribution OSGEO and others have made to providing quality open source applications which enable the OGC standards to be put into practice, as well as many other geospatial applications.

References

AODC (2008) Marine Community Profile of ISO19115, Australian Ocean Data Centre. (last checked 2013)`http://www.aodc.gov.au/files/MarineCommunityProfilev1.4.pdf`

GeoNetwork Opensource. GeoNetwork is a catalog application to manage spatially referenced resources.Ê `http://geonetwork-opensource.org/` (last checked 2013)

GeoServer. GeoServer is an open source software server written in Java that allows users to share and edit geospatial data.`http://geoserver.org` (last checked 2013)

Hollmann, C., Stasch C., Jirka, S., (2010) Coupling 52° North's Sensor Observation Service and Geoserver's Web Feature Service. Presentation at FOSS4G 2010 (last checked 2013) `http://2010.foss4g.org/presentations_show.php?id=3800`

Jasper Reports. `http://www.jaspersoft.com` (last checked 2013)

MapServer. MapServer is an Open Source platform for publishing spatial data and interactive mapping applications to the web. `http://mapserver.org/` (last checked 2013)

Oliver, H., (2012) The Cylc Forecast Suite Metascheduler User Guide. (last checked 2013) `http://cylc.github.com/cylc/CylcUserGuide.pdf`

OpenLayers. OpenLayers is a open source JavaScript library for displaying map data in most modern web browsers. `http://openlayers.org/` (last checked 2013)

Popinet, S. (2003) Gerris: a tree-based adaptive solver for the incompressible Euler equations in complex geometries Journal of Computational Physics, Volume 190, Issue 2, September 2003, Pages 572-600

Quantum GIS. Quantum GIS (QGIS) is a user friendly Open Source Geographic Information System (GIS) licensed under the GNU General Public License. `http://www.qgis.org/`

PostgreSQL. PostgreSQL is an open source object-relational database system. `http://www.postgresql.org` (last checked 2013)

Scott, J., Wheeler, D., Lucas, M., Herz,J., (2011) Open Technology Development (OTD):Lessons Learned & Best Practices for Military Software. `http://www.oss-institute.org/OTD2011/OTD-lessons-learned-military-FinalV1.pdf` (last checked 2013)

SilverStripe CMS. SilverStripe CMS is an open source web content management system. `http://www.silverstripe.org` (last checked 2013)

Wessel, P., Smith, W. H. F., Scharroo, R., Luis, J. F. (2011) The Generic Mapping Tools Technical Reference and Cookbook. (last checked 2013) `http://gmt.soest.hawaii.edu/5/GMT_Docs.pdf`

Spatial Data Infrastructure of the Plurinational State of Bolivia

A free and democratic SDI

by Raul Fernando Molina Rodriguez and Sylvain Lesage

GeoBolivia (Bolivia). `rmolina@geo.gob.bo`

Abstract

The Vice Presidency of the State, with the help of the GeoBolivia project, is building the Spatial Data Infrastructure of the Plurinational State of Bolivia (IDE-EPB by its Spanish initials). The first phase of the project has already been completed. It consisted in implementing an infrastructure and a geoportal that nowadays gives access to the reference geographic information of Bolivia, through WMS, WFS, WCS and CSW services. The project is currently in its second phase dedicated to decentralizing the structure of IDE-EPB and promoting its use throughout the Bolivian State.

The whole platform uses free software and open standards. As a complement, an on-line training module was developed to undertake the transfer of the knowledge the project generated. The main software components used in the SDI are: gvSIG, QGis, uDig as GIS desktop clients; PostGreSQL and PostGIS as geographic database management system; geOrchestra as a framework containing the GeoServer map server, the GeoNetwork catalog server and the OpenLayers and Mapfish GIS webclient; MapServer as a map server for generating OpenStreetMap tiles; Debian as operating system; Apache and Tomcat as web servers.

Keywords: SDI, Bolivia, GIS, free software, catalog, gvSIG, QGIS, uDig, geOrchestra, OpenLayers, Mapfish, GeoNetwork, MapServer, GeoServer, OGC, WFS, WMS, WCS, CSW, WMC.

Introduction

The Spatial Data Infrastructure (SDI) of the Plurinational State of Bolivia IDE-EPB, created through the GeoBolivia project, aims at providing useful geographic information (GI) to institutions and citizen. The SDI provides relevant, harmonized, and high quality GI, allowing to support the formulation, evaluation, implementation and monitoring of practical policies, with a direct or indirect impact on the territory. Additionally the SDI brings support in terms of education and research.

The GeoBolivia project is responsible of organizing the cooperation with and between all the institutions involved in the field of GI in Bolivia. Indeed, the contribution of public and private institutions and professionals is crucial for strengthening and improving the data.

With respect to these main objectives, the specific goals of the first phase of the GeoBolivia project were to build a server infrastructure, publish GI using open standards[1] and develop a web portal for consultation, as a pioneer and main node of the SDI (Nebert, 2004).

Regarding the selection of technology, the GeoBolivia project opted for the use of free software and OGC standards, ensuring its sustainability and promoting the democratization of access to information. On one side, this decision was based on the conviction that democracy depends on an informed citizenry with a wide access to public information, allowing them to fully participate in public life, collaborate in determining priorities for public spending and fight against corruption through demand accountability (Neuman, 2002). On the other side, the selection of free software was based on the fact that open source geographic software has reached yet a mature technological level that allows to use them exclusively. The use of free software is also endorsed by Article 77 of the new Bolivian General Telecommunications Law that states: "The Executive, Legislative, Judicial and Electoral bodies at all levels shall promote and prioritize the use of free software and open standards, within the framework of sovereignty and national security".

Project Development

Main SDI framework

Initially, various technological options for imple-

[1] The only exception to the rule of open standards is the use of ISO 19139 profile for metadata (see `http://www.opengeospatial.org/ogc/faq/openness/#2` for definition of an "open standard"). It has been decided the benefits in terms of interoperability of using this *de facto* standard justify the exception.

menting the framework of the IDE-EPB have been evaluated, giving priority to free software, modular architecture and community development. In this sense the following solutions were considered: GeoNode (`http://geonode.org/`), EasySDI (`http://www.easysdi.org/`) and finally geOrchestra (`http://www.georchestra.org/`). Although the basic software modules of the three solutions are the same (GeoNetwork and GeoServer), the latter has been chosen principally for its independence to any leading company or international institution, and for the better functionalities of its map viewer module.

Complete software architecture

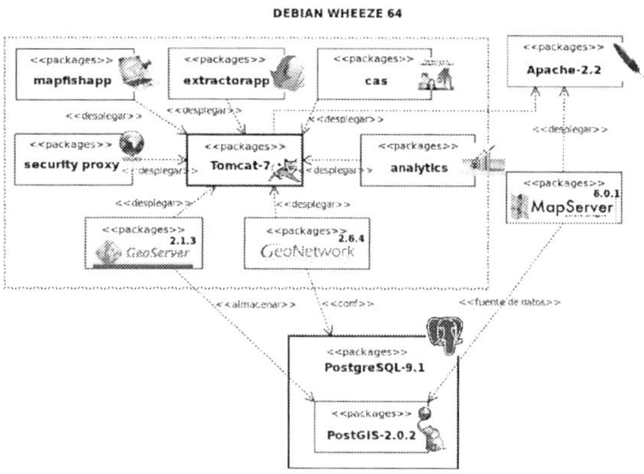

The IDE-EPB software architecture is based on geOrchestra for the management of GI, but also includes other components for all side infrastructure. In every part of the project, free software is exclusively used.

All desktop and laptop computers are using distributions based on Linux. FOSS GIS clients (QGIS, uDig and gvSIG) are used for data processing before storage (encoding, projection, conversion) and creation of styles and symbols for WMS maps. Some functions of the GDAL library are directly used for raster conversion.

The servers use GNU/Linux operating system (Debian Wheezy), Apache2 as a proxy and web server and Apache Tomcat as a Java applications server to deploy the geOrchestra modules.

The geographic components of the SDI, based on geOrchestra modules, are the following:

- A repository of geographic files (SHP and GeoTIFF formats).
- A geographic database (PostgreSQL and PostGIS).
- A map server for generating web services (GeoServer).
- A metadata catalog linked to the web services (GeoNetwork).
- A map viewer with basic GIS client functionalities (Mapfish, based on OpenLayers).
- An extractor, related to the map viewer, for downloading at once various layers of data for a given geographic extent (Extractorapp).
- A map server dedicated to the generation of base maps using OpenStreetMap data[2] (MapServer).
- A central authentication system (CAS), a security proxy and a LDAP users directory, providing shared rights management between all modules.
- An analytics module for monitoring the use of web services.

Moreover, the website and the blog are based on the SPIP content management system[3], documentation and project management are managed by Chiliproject (`https://www.chiliproject.org/`), and the knowledge transfer systems are Moodle (`https://moodle.org/`) for on-line training and OpenMeetings (`http://openmeetings.apache.org/`) for web conferences.

Cycles of development

The SDI implementation was realized following cycles of development, beginning with an initial prototype, and improving it through new versions regularly published in 2012 and 2013. Five versions were published:

Version 0.1: Prototype platform with a simplified installation of geOrchestra and a simple website;

Version 0.2: Installation of the main modules of geOrchestra (map server, catalog, user management), generation of a WMS base map using OpenStreetMap data, update of the website;

[2] As mentioned in the OpenStreetMap wiki, "OpenStreetMap data is free for everyone to use. Our tile servers are not." The IDE-EPB decided to generate its own tiles in order to contribute to the distribution of OpenStreetMap base maps of Bolivia, instead of penalizing community servers by just using their WMS services. See `http://wiki.openstreetmap.org/wiki/Tile_usage_policy`.

[3] `http://www.spip.net/`

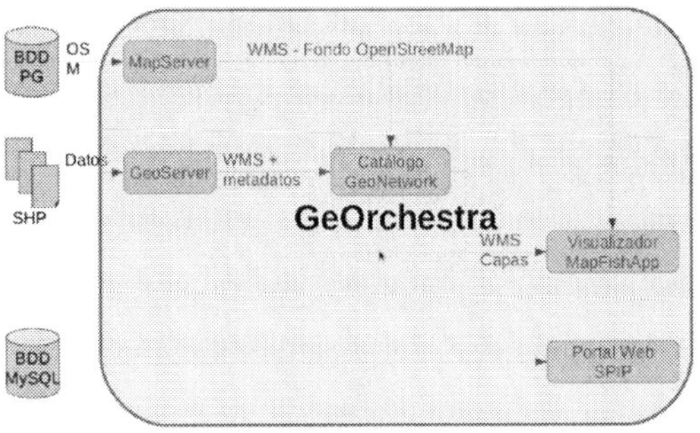

Version 0.3: Installation of a geOrchestra auxiliary module (extractor allowing data download), development of an "API" which enables users to display GeoBolivia data on their own website, new website presentation with a large section providing texts, images and videos dedicated to help users;

Version 0.4: Two SDI nodes (IDE-EPB node and risks management node) interconnected via catalog harvesting (CSW) and maps services (WMS), installation of an auxiliary module of geOrchestra (simple web catalog client), map viewer for mobile devices, Live DVD (based on Ubuntu 12.10) containing all main GeoBolivia data, metadata and styles, and free GIS client tools;

Version 1.0: Creation of "virtual nodes" for institutional partners, hosted on GeoBolivia servers, with remote access and rights to upload, edit and publish their own data, higher level servers infrastructure, with an increased control on the availability of services and data quality.

The following versions are intended to include:

- more contributions to the geOrchestra community and development,
- a webpage for direct download of full datasets,
- community assessment on data and metadata quality, with a comments and ratings system,
- a personal space for every user,
- interconnection with generic opendata catalogs.

Interactions with the geOrchestra community

As a reciprocity to the use of the geOrchestra framework, the GeoBolivia team became an active member of the community by contributing to improvements as for example the internationalization of various modules[4] and animating the community forums. As a result of these interactions, the version 12.11 of geOrchestra (`http://blog.georchestra.org/post/2012/12/16/geOrchestra-12.11-Bolivia-is-out`) was called "Bolivia", and geOrchestra was given a logo representing a condor, the main symbol of the Andean countries. Moreover, various meetings were organized, in Bolivia and France, between the GeoBolivia project team and members of the French community.

Functionalities of the portal

The detailed functionalities of the GeoBolivia portal are the following:

- **Map viewer**
  - Movement, zoom on the map.
  - Add new layers from the catalog or a remote WMS.
  - Change and save the style of a local or remote WMS layer.
  - Query a layer and filter the data displayed.
  - Save or restore a list of layers using Web Map Context (WMC) files.
  - Print a map in PDF format.
  - Display global coordinates (EPSG 4326).
- **Catalog**
  - ISO 19139 metadata profile.
  - Allow to harvest metadata records on distant catalogs.
  - Allow distant catalogs to harvest the GeoBolivia metadata records.
- **Extractor**
  - Possibility to choose an area of extraction.
  - Handle of various export formats (ECW, GeoTIFF, SHP).
  - Rights access management.
  - Sending an email when extraction is complete.
- **GeoPublisher**
  - Allows publishing geographic data from the GeoNetwork catalog, after loading the data in the metadata record.

[4]geOrchestra is now available in English, Spanish and French.

Workflow of data publication

One of the tasks of the GeoBolivia project has been to collect, order and publish GI. The first step was to inventory the existing GI generated or owned by the State. Then, data were formally requested to all public institutions, under the premise that all information generated from State resources must be public and shared. The requested format for vectorial data is Shapefile (.shp) and GeoTIFF for raster data. In case of receiving GI under another format, files are converted using QGIS, gvSIG, uDig or in case of raster conversion, a GDAL library function is used.

Some processing is done using uDig or QGIS in order to check the topology, projection and attributes of the data. Once processed, the GI (vectorial or raster) is stored on the server and published after a step of configuration of the OGC services. Those web services are organized by namespace in the map server, using the name of the GI producer[5].

Each layer is assigned a style (SLD), generally created with the producer of the GI using uDig, in order to display the maps (WMS) with a meaningful symbology.

Finally, an ISO 19139 metadata record is stored in GeoNetwork and linked to the OGC services metadata (WMS, WFS).

Phase I: technological platform and first interactions

The main objective of the GeoBolivia project, in its first phase, is to standardize, and distribute GIS democratize the Plurinational State of Bolivia.

In this sense, a strong emphasis was done on the technological implementation: install and configure a metadata catalog (`http://geo.gob.bo/geonetwork/`), store, organize and publish data and services (`http://geo.gob.bo/geoserver/`), create a central access point to Bolivian GI with a web portal, that links to the viewer, download and catalog client modules (`http://geo.gob.bo/`). In a spirit of transparency of the project process, a shared development platform was installed, enabling to document the entire project development in a systematic and ordered way (`http://intranet.geo.gob.bo/proyectos/`).

Similarly, it is important to note that during the first phase began an institutional coordination, through activities with users and providers of data within the Inter-institutional Committee. Four meetings of this committee were organized in 2012 and 2013, achieving the composition of its members, interactions around the first developments of the IDE-EPB, and the redaction of a draft of Supreme Decree and Regulations for the constitution of this committee. The main objectives of the Inter-institutional Committee (`https://geo.gob.bo/?-El-Comite-Interinstitucional-`) are to define working directions based on the requirements of its members and to define rules and bring support for GI production and publication, in order to foster the interoperability of the information within the State.

As another part of the project, a virtual learning module (`http://geo.gob.bo/cursos/`) has been developed with the purpose of training the users and producers to apprehend new geographic technologies and adopt the IDE-EPB in their workflows.

The GeoBolivia project also promoted the elaboration of Bolivian standards regarding the handling of GI, within a working group composed of public institutions and headed by the Bolivian Institute of Standardization and Quality (IBNORCA). The spirit of the redaction of these standards is to adapt and simplify the ISO:19100 norms, in order to fit the Bolivian reality.

An important element of the GeoBolivia strategy is the nodal vision, that aims at organizing the SDI as a network of thematic or hierarchic nodes, responsible for its own part of the national data, and linked together using web services (OGC). This should provide technical and organizational robustness and sustainability to the SDI. Therefore, in order to facilitate the implementation of new nodes, the GeoBolivia project is working on a Live DVD offering not only a copy of the GeoBolivia database, but also the ability to install all the software applications that GeoBolivia infrastructure uses. It will include geOrchestra in its internationalized version, allowing to replicate the functionalities of GeoBolivia on any institutional server. Nevertheless, the nodes are free to use any other solution, under condition of giving access to the GI through OGC standards. For example, the first thematic node directly integrated to the IDE-EPB nodal architecture (the GeoSinager portal (`http://geosinager.defensacivil.gob.bo/`), dedicated to "risks management") uses the GeoNode SDI framework without the need of any adaptation.

At the end of the first phase of the GeoBolivia project, the IDE-EPB is still mainly centralized, play-

[5]For example, maps generated with GI from the Department of Education are published in a WMS service accessible at `http://geo.gob.bo/geoserver/mde/wms` ("mde" is for "Ministerio de Educación")

ing the role of repository of all the GI recollected from State institutions and giving free access to it with the appropriate metadata.

Phase II: a public SDI, shared and financially sustainable

Whereas the proposal of the first phase of the GeoBolivia project was only to build a central node for managing State GI, the second phase aims at decentralizing the architecture of the IDE-EPB, recycling existing systems and integrating new platforms into a nodal structure. The SDI is intended to evolve from the current structure implemented in the IDE-EPB (a group of servers located in the Vice Presidency) to a nodal structure (multiple servers interconnected among several institutions). This involves articulation and enhancement of thematic information systems which already exist in other institutions. Most of them were the result of much effort and budget, and even if these systems are mostly not SDI nodes, but rather closed systems only dedicated to data consultation, they may be converted into node catalogs of the IDE-EPB.

In the desired nodal structure, the nodes are interconnected, and possibly replicated, using OGC webservices, providing access to all the State GI to the clients independently of the node they consult. The hoped benefits of such an organization are a better robustness to technical or institutional[6] failures, where the loss of a node does not impact strongly the whole structure, and a more precise control of the quality of data thanks to the splitting of GI into expertise thematic nodes.

Another structural challenge to address in the second phase is the integration of GI from players who do not have the resources or technology to build a node, allowing them to manage their data directly in the central node. Such actors may be territorial subdivisions governments, or institutions that only produce few specific data such as the location of thematic infrastructures (banks, schools, projects). To do this, the GeoBolivia project has developed guidelines and tools to allow the creation of GI and also promotes GI exchange protocols to ensure interoperability through standards. This means that each institution according to its competence may create and update remotely its own metadata and data without the need of operating a node by themselves.

In the same goal of associating State actors to the IDE-EPB, the GeoBolivia project will develop a strategy of participatory quality control of information, facilitating the integration of comments and evaluations by data users directly in the metadata catalog and creating a consultative commission of users and providers, that could emit recommendations to the Inter-institutional Committee.

Institutional progress is essential to carry out any project of SDI. The GeoBolivia project in its second phase is promoting the socialization and coordination with various institutions conducting to strengthening the links established with the main entities producing GI in Bolivia, such as: the Military Geographic Institute (IGM), Unique System of Earth Information (SUNIT), National System of Risks Management (SINAGER) and the National Institute of Statistics (INE), since they cover almost sixty percents of Bolivian State data, and therefore their active participation is required for the Inter-institutional Committee.

Furthermore, the close collaboration with international entities such as the University of Geneva, GéoBretagne, geOrchestra, GeoNode and international cooperation will be promoted in order to foster academic activities and specific programming developments, and to coordinate tasks in regard to the management of GI. The inclusion of the Bolivian universities within the activities of the GeoBolivia project will continue similarly to the first phase, offering internships, promoting the use of open standards and encouraging the generation of new information technology. Currently the GeoBolivia project works with the various universities such as UMSA (La Paz), UPEA (El Alto), UATF (Potosí), UMSS (Cochabamba), USFX (Chuquisaca) and UAGRM (Santa Cruz). The second phase will focus on support for the training of new professionals in the area of SDIs.

Conclusions

Whereas the GeoBolivia project is still in progress, some lessons have already been learned, helping to better focus the future efforts. The main experience has been to understand that building a SDI is basically an organizational challenge, in which the technological implementation covers less impor-

[6] As described in Rodríguez-Carmona 2009, a strong tendency in the past thirty years in Bolivia has been to rule the State using a logic of projects, without a general political vision, leading to duplication of efforts and discontinuity in funding. This paradigm is still present in various emerging SDI nodes in Bolivia, but the decentralized structure is conceived to integrate these fragile projects and their data without endangering the whole organization.

tance than the correct comprehension of the social reality of the actors that compose the State.

The IDE-EPB aims at improving the quality of GI circulation within and between institutions of the State. The way the GeoBolivia project choose to achieve it is making an appropriate technological mitigation between efficiency and simplicity. For example, geographic webservices are still largely unknown to the Bolivian geographers and system engineers but are a necessary element of any SDI. In this case, the solution has been to provide an extensive offer of technical training to upgrade the skills of the State technical manpower. On the other end, some public institutions were installed a cutting-edge SDI framework but never adopted it, leaving the database empty. The solution in this case has been to give them a remote access to their data on the central node of the IDE-EPB, instead of operating their proper node, that was overkilling their needs and capacity.

References

González, V., Peñarrubia, F.,Higón, J.,Sanz J., and Anguix, A., 2009, "Spatial Data Infrastructure of Venezuela, a SDI 100% free software", Fourth Conference on GIS FREE, University of Girona, 1-7.

D.D. Nebert, 2004, "Developing spatial data infrastructures : The SDI cookbook", Global Spatial Data Infrastructure, version 2.0.

2011, "Project: Spatial Data Infrastructure of the Plurinational State of Bolivia: Strategy and Technical Document", Vice Presidency of the Plurinational State of Bolivia.http://geo.gob.bo/IMG/pdf/ide_-_vpep_1_.pdf

Rodríguez-Carmona A., 2009, "El proyectorado: Bolivia tras 20 años de ayuda externa", Plural editores / INTERMON OXFAM.

GIS for All: Exploring the Barriers and Opportunities for Underexploited GIS Applications

by Hao Ye[1], *Michael Brown*[2], *and Jenny Harding*[3]

1: Nottingham Geospatial Institute, University of Nottingham; 2: Horizon Digital Economy Institute, University of Nottingham; 3: Ordnance Survey Southampton.
isxhy@hotmail.com

Abstract

Geographical Information Systems have been existed since the early 1960s, but evidence suggests that adoption of GIS technologies still remains relatively low in many sectors. We will explore both the barriers that affect the utilisation of GIS and opportunities to overcome these barriers. As part of this exploration we performed a literature review, collected responses from quantitative questionnaire survey and interviewed a range of technical and domain experts. Having analysed and collated the results of these studies we have identified ways forward for future research and development to facilitate wider spread adoption and exploitation of GIS applications. Our discussion focuses on the importance of open-source GIS software, open data and cloud computing as key mediators for breaking the barriers and promoting the wider appropriation of GIS based solutions.

Keywords: GIS barriers, GIS opportunities, Open-Source GIS, Open Data, OpenStreetMap, Cloud Computing.

1. Introduction

The development of Geographical Information Systems (GIS) began in the early 1960s and rapidly advanced since the late 1980s. Over the past 50 years, GIS technology has been increasingly introduced to a wide range of sectors. In addition to planning agencies and local governments, many other sectors have been involved, such as social science, transportation, earth science, military, agriculture, environmental protection, etc. The integration of GIS technology in traditional geospatial tasks provides a number of optimal solutions for individuals/groups, e.g. policy developers, decision-makers, managers, researchers, and allows the performance of existing tasks cheaper, faster and more completely (Elangovan 2006).

The multidisciplinary nature of GIS technology means that the diffusion, appropriation and use of GIS technologies are distributed in a variety of subject domains, was and is often regarded as a new technology or approach on traditional task with specific disciplines. This characteristic means that the adoption of GIS technologies have not yet fully delivered its potential or adequately addressed the need of GIS users in many domains (Ventura 1995, Nedovic-Budic 1999). Early studies have mentioned that most users adopt GIS technology in relatively simple applications but rarely use it for sophisticated tasks (Nedovic-Budic 1998). More specifically, these studies indicated that the expectation of GIS technology in the areas of geospatial analysis and decision-making was particularly high but the penetration is relatively low. The task demands were either not achieved at all or not achieved as expected due to barriers and factors that hinder the utilisation of GIS.

Some researchers have investigated the direct or indirect barriers which affect the adoption rate of GIS from early stages (Ventura 1995, Nedovic-Budic 1999, Esnard 2007, Akingbade et al. 2009). This previous work has identified the existing barriers found for effective use of GIS and classified them into two board groups(Brown 1996, Brown & Brudney 1998, Esnard 2007, Yap et al. 2008, Asligui Gocmen & Venturac 2012) ;

- **Organisational barriers**: generally referred to department factors, such as lack of staff (e.g. constraints by size of the team or funding), lack of purpose or mission to promote GIS application and lack of collaborators and networking.
- **Technical barriers:** including lack of context, insufficient software and tools, lack of reliable data and lack of technical knowledge.

In addition, another difficulty is the fast paced updating and development of GIS technology. Empirical knowledge is still scarce on the changing barriers accompanied with rapid advances of GIS technology, as well as potential solutions to justify the possibility for problem-solving.

The aim of this paper is to gather information from the literature, as well as questionnaire responses and expert opinions to identify and update the GIS barriers. In addition we aim to explore the potential opportunities, in terms of Open-Source GIS, Open-Data and Cloud Computing in overcoming these barriers. The current application domains of GIS and existing barriers were re-examined as it is important as focal points for the identification of future opportunity and proposal of potential solutions. This study intends to explore the following research questions:

1. Who are the current users of GIS and what is the current state of GIS applications?
2. What current opportunities exist for exploitation of GIS beyond commonly accepted applications?
3. What are the existing barriers that result in the underutilisation of GIS?
4. What possible solutions exist for overcoming the barriers and realising the opportunities that have been identified?

This study provides a brief overview of the inroads made by GIS, highlight a way forward for GIS researchers, professionals and developers to fulfil potential users' unmet needs and promote ever-changing practical development. The rest of this paper is divided into 4 sections. Section 2 gives an overview and analysis of literatures exploring the current application and barriers of GIS in several selected domains. Next, Section 3 describes the results of a questionnaire survey and interviews with GIS users and experts exploring this area. Section 4 discusses the potentials of recent technologies toward breaking the barriers. Finally section 5 provides general conclusions.

2. Literature Survey: Applications and Identified Barriers

In order to have an impression of the current state of GIS applications, as well as an update of the impediments for GIS adoption, literature was searched and revisited, and analysis was conducted for application and barriers identified in past studies. In this paper the literature review focused on 4 well-known fields of GIS application, including *Government, Transportation, Commercial Sectors, Public and Educations*. The selection might not contain comprehensive application domains, but it draws an impression about the trends of the mainstream GIS applications. The literature search for relevant publications was performed by using web-based search engine. The search was carried out based on a Boolean search containing the keywords such as "GIS barriers", or "GIS" and relevant 'application field' selected above. The searched articles and referenced articles containing the terms were selected and reviewed. The GIS application domains and identified barriers were analysed and summarised.

Government

The government sector was an early adopter of GIS, and has been a major user over the last 20 years. GIS has been utilised in major governmental sectors such as land management, coastal condition monitoring, census statistics, disaster management, environment protection, military, administrative purposes, surveying, etc (Brown & Brudney 1998, Longley et al. 1999, Montoya 2003, Fleming et al. 2009, Sunarto 2011, Gregory 2012, Mendas & Delali 2012).

The review revealed that when GIS emerged as a valuable technology in the 1990s, a number of researchers started to examine the barriers of GIS implementation in local governments (Brown 1996, Nedovic-Budic & Godschalk 1996, Brown & Brudney 1998,). These studies primarily identified three types of main barriers for government staffs, including: (1) lack of programs/software for desired application; (2) lack of awareness of available tools; (3) lack of national initiatives. In recent years, some extensive studies have been conducted to reveal that, except for former barriers, some updated barriers for GIS users. For example, current GIS users cannot catch up with the technology advancements. One typical example is the adoption of emerging internet-based GIS in urban planning and decision-making (Peng 2001, Drummond & French, 2008). Some other studies suggest that the barriers of GIS adoption in planning sectors are similar to the impediment of adoption of GIS in other governmental sectors (Greetman & Stillwell, 2004).

Transportation

Transportation has always been one of the most important domains of GIS application. To date, GIS has been applied to different transport problems, such as short path analysis, vehicle routing, trip investigation, road navigation, dynamic routing, accident analysis, Intelligent Transport System (ITS) (Aultman 1998, Xu 2005, Pons & Perez 2003, Kim et al

2011, Deshpande et al 2011).

Past studies have revealed that transport modellers are used to coding up abstract road networks and managing transportation data non-spatially. Many transportation planners are unfamiliar with GIS data management and tools. Therefore, at that time the barriers to GIS utilization are likely to be technical, referred to as lack of awareness. Other barriers identified included insufficient tools and data required by many transportation applications (Sutton 1996). Researcher in highway sectors summarised the barriers for implementing GIS application include: (1) unaware the benefits of GIS; (2) lack of good communication between highway engineers and GIS engineers; (3) standards for data consistency; (4) insufficient data; (5) lack of knowledge in understanding GIS methods (Smith 2012).

Commercial Sectors

Since GIS contains of many powerful software toolsets for the integration of commercial data and geographical information, it enables a wide range of opportunities for different business domains, such as banking, insurance, logistics, media, real estate and retail (Jafrullah 2003). A detailed summary has been provided by about the application of GIS in commercial sectors, as well as the techniques that could be adopted for the applications by Dangermond (2001). The primary use of GIS in commercial sectors is to analyse the distribution of customers in order to expand customer base, improving service quality and consistent business profitability. In fact, the "Geo-marketing" is a term dedicated to describe the commercial GIS, which include the studies of consumer's spatial behaviours, marketing and consumer's values, lifestyle and geographical information as well as the location of retail for market demand and service activities (ESRI 2003, Gerard 2006).

The past studies shown that the main barriers of GIS in social sectors (including commercial sectors and non-profit sectors) are the cost of GIS and lack of internal expertise. Finance resources were insufficient to support in-house GIS at market value and outsourcing costs exceeding $100 per hour are not easily justified for small and medium organisations. It also mentioned that the increase of GIS application practices is important for the consideration of potential selection of GIS in various sectors (Edwards, 2011).

Public & Education

GIS have attracted wide interests for public and education sectors in the past 30 years, e.g. measuring libraries accessibility, aquariums, building attack simulation, high schools and university education (Wheeler et al. 2010, Park 2012, Kevany 2003, Singh et al., 2012).

The research of integrating GIS as tools into school curriculums have been conducted for a deeper-learning experience and participation. However, three main challenges were identified as: (1) hardware issue (i.e. powerful computing to store and interpret data); (2) software and data (i.e. free is important for school); (3) teaching skills and competence (Wheeler, et al. 2010). A study conducted by Read et al. (2005) has highlighted the benefits and potentials of adopting GIS for educational marketing, but a few significant barriers referred to general implementation were also addressed, such as: (1) lack of data and skill exchange and cooperation between department; (2) lack of user-friendly software interface for novice users; (3) acceptance of new tools and methods for traditional tasks; (4) expensive geo-demographic data; (5) licence permission for GIS software. This study particularly draw attention to insufficient GIS data sources, especially these high-quality and well-documented data.

Table 1: The review summary of GIS application domain and the identified barriers.

Domain	Application areas	Identified barriers
Government	• *land management* • *coastal monitoring* • *environment protection* • *census statistics* • *fire management* • *military* • *administration* • *surveying*	• *insufficient programs* • *lack of awareness of tools* • *lack of initiatives/mandates* • *lack of support from managers* • *understanding technology*
Transportation	• *short path analysis* • *vehicle routing* • *trip investigation* • *road navigation* • *dynamic routing* • *accident analysis*	• *insufficient tools* • *lack of expertise* • *lack of professional staffs* • *training* • *understanding technology* • *lack of communication*
Commercial Sectors	• *banking* • *insurance* • *logistics* • *media* • *real estate* • *retail*	• *expensive cost* • *lack of expertise* • *lack of data*
Public & Education	• *libraries* • *museums* • *nature centres* • *high schools* • *Universities*	• *lack of cooperation* • *insufficient software and data resources* • *lack of proficient teachers* • *acceptance of new technology*

The survey in this section has conducted a condensed review of literatures about the GIS application domains, as well as the barriers for improving GIS adoption rate. It is clear, from the summary in Table 1, that the applications domains of GIS vary, but barriers in different sectors are not incompatible. If dividing the barriers into organisational and technical factors described above, the government suffer more from organisational factors such as understanding technology, lack of awareness, and lack of organisational support. The Transportation sectors are restricted by some technical factors such as lack of tools and expertise. The commercial sectors expect cheaper and better software, and more reliable and convenient data. Public and Education sectors expects more cooperative and experts. These barriers show different underutilisation and potential expectation from GIS technology.

3. Questionnaire Survey for Barriers of GIS Use

A questionnaire survey was designed to identify the entry barriers from current GIS users and underutilised opportunities from their respective backgrounds. Questionnaires were sent to 30 candidates to inform the survey and the authors collected 22 valid questionnaires from them. These questionnaire participants were mainly selected from two organisations: the school of Civil Engineering at the University of Nottingham (10 candidates, all of them are PhD students, various engineering backgrounds) and Ordnance Survey in Southampton (12 candidates, 8 staffs in research team and 4 intern students, all of them have geo-science backgrounds), see Table 2 for the statistical summary. Before the survey, the authors investigated the candidates' backgrounds and their relevance to GIS. The criterion for screening was that participants needed to have basic knowledge of GIS knowledge, and at least some experience using GIS software.

Table 2: The summary of participants' backgrounds.

Organisation	Percentage	Backgrounds	Percentage
University of Nottingham	46%	*Geospatial Engineering.*	28%
		Transport Engineering.	9%
		Civil Engineering.	9%
Ordnance Survey	56%	*Geoscience*	56%

The survey questionnaires were designed into two sections: the first section asked relevant personal details, e.g. major in university, current departments; the other section includes some multiple-choice questions and open questions, both referred to the investigations to GIS barriers and potentials. All participants were asked to complete questionnaire and taken part in an in-depth interview exploring their responses to open questions.

Quantitative Results

In order to understand the range of expertise represented, we asked participants to report their proficiency with GIS from one to three: 1) Beginner: just know general principles or ideas of GIS and how to use basic GIS functionalities; 2) Trained User: trained formally by GIS degree, GIS course and workshops; 3) Expert: used GIS up to 5 years or more. The results in Figure 1 show that as is to be expected all participants reported some knowledge of GIS, but otherwise there was a relatively even spread of expertise.

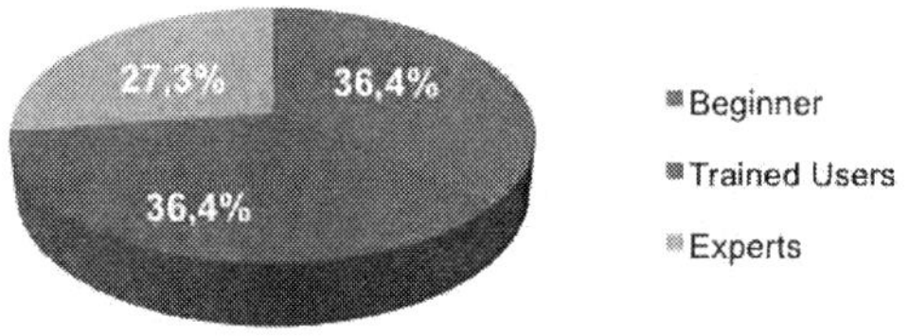

Figure 1: self reported particitipant GIS expertise.

In order to ascertain initial thoughts about barriers to entry, participants were asked to select the three most important barriers that they have encountered from a list of seven identified during the literature review. The results of the response were shown in Figure 2.

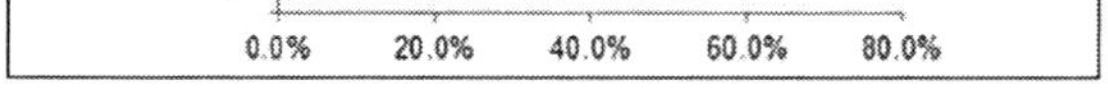

Figure 2: Barriers to entry for GIS .

From the results shown in Figure 2, the most selected factors from participants were *lack of GIS traning* (72.7%), *lack of GIS awareness* (54.5%) and *lack of funding* (45.5%). The three factors can all be seen as organisational factors rather than technical constraints. The development of cheaper, easier to learn GIS could help resolve both lack of GIS lack of funding and training. Surprisingly, the results revealed that many users selected software usability as a significant barrier, rather than lack of data and lack of software. This is an interesting finding as many participants are novice users. This effect may be due to the fact that the first impression of GIS for new users is related to interface or operation menus, which affects their impressions of GIS as a tool.

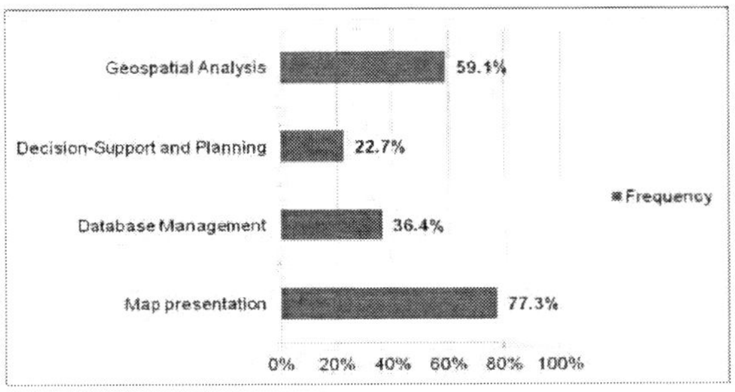

Figure 3: The benefits of GIS technology for users.

The final quantitative aspect of the questionnaire asked particiants to identify the key benefits of using GIS in their work. Four main uses of GIS identifid during the literature reviewwere listed and users were allowed to tick multiple options. The results show that the map presentation, is rated highest (77.3%). In addition, an interesting finding is that the use of GIS for decision-support and planning was only 22.7%. Decision-support and planning using GIS are regarded as promising applications by some but these are apparently still seen as a secondary or future functions by many users. This might be the reason because the participatants are more engineers and researchers rather than decision makers.

Barriers of GIS Use Identified

In-depth semi-structured interviews were performed with domain experts. These experts were selected as they are in close contact with a multiplicity of GIS users from all fields having therefore an extensive knowledge about the problems and needs of GIS users. In this research a total of 6 groups were selected for this interview, including 3 team leaders in Ordnance Survey (i.e. Customer Service, Post-Sales Support, and User Need Research Team) and a technical researcher in Chinese State Bureau of Surveying and Mapping and a renowned academic researcher at University of Nottingham.

Ordnance Survey Customer Service This group noted a range of barriers to GIS use from customer reports as customer service is the main department contacting users, including: (1) methods for data acquisition; (2) data formats and attributes; (3) data manipulation (e.g. download, open, transformation); (4) suitable software for processing data; (5) data specifications modification; (6) data update and maintenance; (7) data quality and reliability. In balance, Customers always hope everything is already there without processing or transformation of anything by them. The participants noted that the majority of customers contacting them use only basic GIS applications. These applications mainly include commercial mapping, web-fit mapping and personal map tracking. They also noted that in many cases people contacting customer service were not using or needing to use the more detailed geographic data or sophisticated software, highlighting demand for easy-to-use and low-cost GIS toolsets free of obstructive technical jargon. Novice users prefer a box of toolkits and data included and easy-to-use, but expert users have higher level demands for map data because they want to modify and update by themselves.

Ordnance Survey Post-Sales Support The Post-Sales team supports Customer Service staffs in resolving specific technical issues, i.e. those relating to software, data or coding. One participant mentioned that the majority of their customers are new users of GIS and that they come from two main sectors: *Commercial* and *Government*. Currently, the main tasks from commercial sector are focused on tracking their commercial assets e.g. water pipelines, insurance properties and transport objects. They estimate that around 50% of the customers are commercial users. The participant revealed that large companies often have their own GIS teams, but most commercial companies lack even a small team. Enquires from the government sector are often associated with the interpretation of new data products. The majority of needs for support concern solving GIS manipulation problems, such as data format transformation. The application of GIS in government sectors varies, but participants found the majority are interested in analysis and routing purposes. For example, using ITN (Integrated Transport Network, a type of Ordnance Survey mapping product) in mapping and analysing footpath networks, efficient fuel consumption and addressing encoding. Some government departments have dedicated GIS teams. These may use GIS for relatively advanced applications such as air pollution assessment. Other government departments might use IT professional or technician to perform GIS analysis.

Ordnance Survey User Needs Researchers Understanding customer's unmet needs is important for developing geographic information products. Through research barriers have been found to vary with the level of expertise of the users. For example, it is maybe relatively difficult for novice users to transform data sets to required formats or understand the attributes within the data. However, for users needing to carry out advanced data analysis, in depth understanding of data specifications, data restructuring and attribute reclassification may be required. Professional knowledge and training can be an issue even at this high level of expertise. Another barrier to use can be insufficient computation power for large GIS data sets. Users who require massive data volumes for national coverage may suffer from low-speed computation in their systems. Modern developments in advanced data acquisition techniques (e.g. sensor data) increases the spatial and temporal density of data collected and is likely to further accentuate this problem.

Technical Researcher This participant shared the opinion that existing GIS software is too complicated for non-expert GIS users. While many GIS companies have made much effort in this direction, they feel that the results to date are not satisfactory. Many simple data viewers and data enquiry tools provide simple functions, such as ArcView, Arc Explorer, but they feel that there is a need for novice friendly GIS software that incorporates more sophisticated GIS functionality. This group suggested two possible solutions to overcome this issue: (1) Develop domain specific GIS solutions for professional users that allow users to complete a limited number of advanced GIS operations; (2) Use data middleware, i.e. an abstraction layer that hides detail about hardware devices or other software from an application in order to provide users with a seamless and simplified interaction.

Geospatial Academic Researcher This research group highlighted the need for GIS experts to make systems that are more closely aligned with user's ways of thinking. According to their comments, "the largest challenge for GIS users is GIS terminology which is regarded as too sophisticated, too technical for user". They have found that with applications such as Google Maps and Google Earth, Google addressed this problem more completely than most expert GIS systems aiming at non-expert users, leading to a move towards web services over dedicated local software, for novice users. In addition to terminology as a barrier, they note that data acquisition should not be ignored given the current cost involved in accruing some data sets. Open Street Map is expected to play an important role for GIS users and developers in future". The participants emphases that training is critical to solve many barriers of GIS, such as lack of awareness and software usability". In response to potential opportunities of GIS, this group identified a wide range of possibilities including In-car applications, real-estate, and simulation. They highlight the integration of GIS and social media (e.g. Facebook, Twitter) to analyse human' behaviours in daily life as an area with huge potential, although it does raise issues such as privacy, security and data ownership. They also mentioned the importance of earth observation including measurements and monitoring of the earth, under water, land surface, air and water quality, atmospheric condition and measures of the health of human, plants and animals.

Identified GIS Barriers Summary

Table 3: The identified barriers from GIS users and expectation for future GIS development.

	identified barriers from participants	expectations on future GIS
Questionnaire Survey	• *lack of fundamental training of GIS (novice users)* • *lack of user-friendly GIS* • *lack of GIS training and meet demands for different level of users* • *lack of knowledge for data processing* • *lack of all-in-one GIS toolbox* • *terminology is to sophisticated and document is too technical* • *lack of integration of GIS with various backgrounds*	• *Develop more training (e.g. workshops, e-learning and online resources)* • *Develop assemble from different teams* • *Develop implementation of ideas* • *Develop user-friendly GIS software and documents*
Expert Knowledge Survey	• *lack of readily available data sources (need transformation)* • *lack of professional teams in many sectors* • *concern data quality, maintenance and reliability* • *lack of specific GIS solutions for professional users* • *lack of flexible modules or GIS data middleware* • *lack of computing power* • *documentation is too technical* • *High performance hardware is too expensive*	• *Introduce users with various data sources and the acquisition method* • *Develop more trainings, avoid technical jargons* • *Develop high reliable and high quality geospatial data* • *Develop software for specific solution* • *Require high-computing GIS* • *Develop more GIS application in new fields (e.g. social media, earth observation)* • *Develop more applications in expert or decision making system*

Based on the results of the questionnaire survey and interviews with field experts, the authors have identified some barriers from these users and their expectations for GIS developments.

- Lack of awareness
- Lack of communication
- Entry cost

- Lack of required software
- Insufficient data sources
- Lacks of computing power
- Usability

4. Potential of Recent Technological Development towards Breaking the Barriers Identified

Having explored the current uses of GIS, identified barriers that prevent its wider appropriation and exposed the potential opportunities; we now discuss the ways to overcome these barriers with relatively recent technical developments.

Open-Source GIS

With the development of Open-Source community, a large number of software tools have been established. Up to present, GIS software and toolkits have published online and allow users to freely download, and has been summarised, see Table 4 (Chen et al. 2010). Open-Source GIS have been developed into a broad range of classification. The main spectrum is shown in Figure 4. Open-Source GIS can be approximately categorised as *Desktop GIS, Spatial DBMS, Server GIS and Web GIS, GIS Library, Database, Spatial Tools, etc.* A detailed classification and information for different types of Open-Source Software has been elaborated by Steiniger & Hunter (2012).

Table 4: The main web sources for Open Source GIS software (from Chen et al. 2010).

No.	Name of sources	URL	Note
1	*Open-Source GIS*	*http://opensourcegis.org/*	over 250 softwware packages
2	*FreeGIS*	*http://freegis.org/*	more than 300 software packages
3	*SourceForge*	*http://web.sourceforge.com/*	all sorts of GIS
4	*OSGeo*	*http://www.osgeo.org/*	a list of popular GIS packages

There are a number of advantages of using open-source GIS software to break the barriers identified in previous literature and questionnaire sections.

The first being of Open-Source GIS is that it is free to potential users. The free Open-Source GIS allows novice users to understand the basics of GIS with very low entry cost, and in some degree, it promotes the perception of GIS adoption rate. In fact, this is particularly valuable for individuals, small research group, small commercial companies, education sectors, with insufficient funding. Training sessions or workshops are easier by using Open-Source GIS if focusing on different level of users, but there is a barrier as most of current Open-Source GIS manuals are centred on the technical aspect of GIS rather than case study demonstration. In addition, the training courses are often as expensive as GIS training provided by commercial software providers.

The second advantage of Open-Source GIS is original codes are accessible so it is convenient for software modification. This is highly desirable by the high-level users who want to develop their applications. However, the quality of the open-source codes sometimes cannot guaranteed, and it also add additional cost (e.g. learning knowledge, correct the code errors, time for coding) . Detailed and understandable document is a good approach to solve this problem, but as stated before, it is too technical of current documents.

The third advantage is Open-Source GIS is all freely available online, which enhanced the accessibility for users from different regions, e.g. developing countries. The installation volume of Open-Source software is less than commercial GIS Desktop software systems if only selected plug-ins are utilised. The advantages of these technical and economic aspects offer potential for using Open-Source GIS in various organisations.

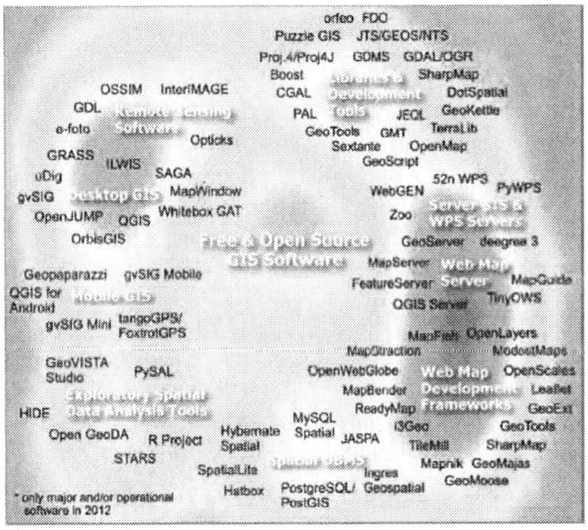

Figure 4: Open-source GIS classification (obtained from Steiniger & Hunter 2012).

Open Data and OpenStreetMap

Geospatial data is at the core of almost all GIS applications. The acquisition of available geospatial data is usually time-consuming, expensive and may be complex. This is mainly due to the disciplinary nature of GIS, as wide range of data is documented by different organisations with different size, format, specification, etc. In recent years many organisations have started to release rich open data for public or developer use in order to promote GIS application and collaboration. For example Open Street Map and Ordnance Survey's OS OpenData™ datasets (OpenStreetMap 2013, OpenData 2013).

OpenStreetMap is a specific open access data set, basically an online web map, created by volunteers to be freely created, shared and delivered to different communities. The map formats satisfy OGC standards so that they could be used in almost all open-source and professional GIS software. The OSM now includes a wide range of data. For some location-based services application, it is more suitable to use OSM data. OSM has advantages over many professionally created data sources, in terms of access to some data that is difficult for professional surveyors to obtain, such as cycling places, city toilets, etc. OSM can be downloaded as different format on for web-mapping or desktop software (e.g. QGIS and ArcGIS). Even some local government departments have started to use OSM for some GIS projects, such as the Birmingham cycling parking project (Birmingham City Council 2012).

However, it was reported that in 2010 the coverage of OSM was up to 69.8% worldwide and this means that the completeness is still lower than existing commercial products (Feldman & Morley 2012). The data quality of OSM is the main concern with respect to different applications because all mappers are volunteers. In addition, the tags seem sometimes not to be quality checked leading to confusion. A high learning curve is necessary to acquire the data and prepare it so it can be reasonably used in own analysis. A recent study reports that the geometrical accuracy of OSM was assessed between 5 and 10 metres, with higher accuracy in urban areas but low accuracy in rural/deprived areas) (Feldman & Morley 2012). These disadvantage limited the wide spread of OSM which might bring potential opportunities if these issues can be solved.

Cloud Computing

Cloud computing is rapidly emerging as a technology for industries who provide software, hardware and infrastructure. In the field of GIS, the introduction of cloud computing for GIS software and services has become a popular topic. There are several types of deployment scenarios, for example, the ESRI have attempted to develop a Cloud-based platform for Amazon (ESRI 2012). One critical difference between GIS in cloud and traditional GIS is that it could deliver GIS services via the internet, instead of installing and running the software on personal computers. Service users connect to private or public clouds (a set of connected computing computers) and request service from cloud providers (Kouyoumjian 2010). Due to completely different service mode, the future cloud GIS might offer many benefits and some of them might break the barriers of current barriers. The summary in Table 5 organises a set of potential advantages of GIS, as well as the analysis of their potential to break the barriers of GIS.

Table 5: Summary of potential benefits and issues of Cloud GIS.

Cloud features	Potentials of breaking barriers	Potential Issues
shared GIS resources with multiple users	• *promote the GIS adoption, training and information exchange*	• *data leak and privacy*
lower total cost	• *reduce entry (development) cost*	• *still need cost for users*
pay-as-you-go computation	• *reduce entry cost for different users*	• *uncertain benefits for small organisation*
high-speed cloud server	• *satisfy high-speed computation demands*	• *high cost* • *data privacy* • *reliability*

Cloud GIS indeed can potentially provide new benefits based on traditional GIS, such as promoting productivity and reduce cost. Particularly, as the identified barrier of high-computing requirement for professional users, the advantage of cloud computing can help meet this demand are hitting the ceiling of hardware cost, storage space and computation resources (Willoughby 2011). However, some potential issues will be promoted; for example, it is difficult to judge whether it is safe to send data (e.g. sensitive data) onto a cloud server.

5. Conclusions

The research in this paper explored both the barriers that affect the utilisation of GIS and opportunities to overcome these barriers. The research provides a literature review related to the application domains of GIS, as well as the identified barriers. Next, a survey analysis from a pool of selected participants was carried out. The results from survey respondents showed that the applications of GIS still confront many barriers, including: (1) lack of awareness; (2) lack of communication; (3) entry cost; (4) lack of required software; (5) insufficient data sources;(6) lacks of computing power; (7) usability. In addition, many applications are still focused on traditional map presentation and geometry and topology based querying, decision-support and database management as a high-level functions are still under-utilised.

The research discussed a set of potential solutions to break the barriers and create potential opportunities for GIS development. For example, the development of Open-Source GIS offers a set of rich software toolsets in many different fields that could contribute efficiency and cost-effectiveness for these challenges; such as reducing entry cost, improve accessibility, provide more tools. Open data potentially offers users with free and easy-to-use data. This research addressed OpenStreetMap as a rich data source, but the information completeness is low compared to some commercial products and quality of data is a concern that must be considered. The Cloud Computing potentially provides on-demand and high performance computation. However, it is sensitive to send data to the servers as it causes privacy issue of data owners. In balance, this research provides some ideas to put forward these valuable resources for developing potential GIS solutions in these identified domains.

Acknowledgement This work was carried out in collaboration with Horizon Digital Economy Research, the University of Nottingham, through the support of RCUK grant EP/G065802/1. We would appreciate the contributions research team of all survey participants during entire research. We would also appreciate the comments and feedback provided by the reviewers which were extremely helpful in improving the quality of this paper.

References

Akingbade, A., Navarra, D. D., Georgiadou, Y. (2009). 'A 10 year review and classification of the geographical information system impact literature 1998-2008', *Nordic Journal of Surveying and Real-estate Research*, 4, 84-116.

Asligui Gocmen, Z. & Ventura, J. S. (2012), 'Barriers to GIS use in planning', *Journal of the American Planning Association* 76 (2), 172-183.

Aultman, H. L., Hall F. L., Baetz, B. B. (1998). 'Analysis of bicycle commuter routes using geographic information systems', *Transportation research record* 1578, 102-110.

Birmingham City Council (last checked 2012) 'Bike North Birmingham - Transforming walking and cycling in North Birmingham'. `http://www.birmingham.gov.uk/bikenorthbirmingham`

Brown, M. M. (1996), 'An emprical assessment of the hurdles to geographical information system success in local government', *State and Local Government Review* 28 (3), 193-204.

Brown, M. M. & C. J. Brudney (1998), 'A smarter, better, faster and cheaper government: constracting and geographical information systems', *Public Administration Review*, 58 (4), 335-345.

Clapp, J. (1998). 'Using a GIS for real estate market analysis the problem of spatially aggregated data', *Journal of Real Estate Research*, 16(1), 35-56.

Dangermond, J. (2001), 'Geographic information systems and science – The commercial setting of GIS', Wiley Press, 55-65.

Deshpande, N., Chanda, I., Arkatkar, S.S. (2011), 'Accident mapping and analysis using geographical information systems', *International Journal of Earth Sciences and Engineering*, 4, 342-345.

Drummond, W.J.& French, S. P. (2008), 'The future of GIS in planning: converging technologies and diverging interests', *Journal of the American Planning Association*, 74(2), 161-173.

Edwards, S. J. (2011), 'Geographical information system in the social sector: trend, opportunities and barriers and best practices', Master Thesis, Greensboro, University of North Carolina.

Elangovan, K. (2006), *GIS: Fundamentall, application and implementation*, New India Publishing Agency, Jai Bharat Printing Press.

Esnard, A. M. (2007), 'Institutional and organisational barriers to effective use of GIS by community-based organisations', URISA Journals, 19 (2), 14-21

ESRI (last checked 2012), 'Evaluate ArcGIS for Server on Amazon Web Services' URL. `http://www.esri.com/software/arcgis/arcgisserver/evaluation`

ESRI (2007), 'Enterprise GIS for Utilities – Transforming Insights into Results', ESRI White Paper, J9658, 1-7.

Feldman, S. & Morley, J. (2012), 'Open Street Map -GB Overview', 2012 *Open-Source GIS Conference*, University of Nottingham, UK.

Fleming, S., Jordan, T., Madden, M., Usery, E.L., Welch, R. (2009), 'GIS applications for military operations in coastal zones', *ISPRS Journal of Photogrammetry and Remote Sensing* 64, 213-222.

Gerard, G. (2006), 'Geo-marketing: method and strategies in spatial marketing', ISTE Ltd, ISTE USA Press.

Greetman, S. & Stillwell, J. (2004), 'Planning support system: an inventory of current practice', *Computers, Environment and Urban Systems*, 28 (4), 291-310

Gregory, I. N. (2002). 'The accuracy of areal interpolation techniques: standardising 19th and 20th century census data to allow long-term comparison', *Computer, Environment and Urban System* 26(4), 293-314.

Jafrullah, M., Uppuluri, S., Rajopadhaye, N., Reddy, V.S (2003). 'An integrated approach for banking GIS', *Map India Conference – Business GIS.*

Kevany, M. J. (2003), 'GIS in the World Trade Centre attack – trial by fire', *Computers, Environment and Urban Systems* 27, 571-583.

Kim, M., Miller, E. H., Nair, R. (2011), 'A geographic information system-based real-time decision support framework for routing vehicles carrying hazardous materials', *Journal of Intelligent Transportation System: Technology, Planning and Operations* 15(1) 28-41.

Kouyoumjian, V. (2010), 'GIS in the Cloud, the new age of cloud computing and geographical information system', *ArcWath e-Magazine*. ESRI.

Lam, C. C., Lai, E., Wong, J. (2009) 'Implementation of geographic information system (GIS) in secondary geography curriculum in Hong Kong: current situation and future direction', *International Research in Geographic and Environment Education* 18 (1), 57-74.

Longley, P.A., Goodchild, M. E., Maguire, D. J., Rhind, D. W. (1999), *Geographical Information Systems: Principles and Applications*, 2nd Edition edition, John Wiley & Sons.

Mendas, A. & Delali, A. (2012), 'Integration of Multi-criteria Decision Analysis in GIS to develop land suitability for agriculture: Application to durum wheat cultivation in the region of Mleta in Algeria', *Computer and Electronics in Agriculture* 83 117-126.

Montoya, L. (2012), 'Geo-data acquisition through mobile GIS and digital video: an urban disaster management perspective', *Environmental Modelling & Software* 18, 869-876.

Nedovic-Budic, Z. & Godschalk, D. (1996), Human factors in adopting of geographical information systems: a local government case study: Public Administration Review, 55(6), 554-567.

Nedovic-Budic, Z. (1998). 'The impact of GIS technology', *Environment and Planning B: Planning and Design* 25(5), 681-692.

Nedovic-Budic, Z. (1999), 'Evaluating the effects of GIS technology: review of methods', *Journal of Planning Literature* 13, 284-295.

OpenStreetMap Team (last checked 2013), 'OpenStreetMap: the free wiki world map', URL. `http://www.openstreetmap.org/`.

Ordnance Survey Open Data Team (last checked 2013), 'Business - OS OpenData', `www.ordnancesurvey.co.uk/oswebsite/products/os-opendata.html`

Park, S. J. (2012), 'Measuring public library accessibility: a case study using GIS', *Library & Information Science Research* 34, 13-21.

Peng, Z. R. (2001), 'Internet GIS for public participation', *Environment and Planning B*, 28(6),889-905.

Pons, J. M. S & Perez, M. R. (2003), 'Geographical information systems and intelligent transport systems technologies used to form new communication networks', *Network and Communication Studies* 17, 53-70.

Read, P., Higgs, G., Taylor, G. (2005), 'The potential and barriers to the use of geographical information systems for marketing applications in higher educational institutes', *Modelling Intelligence & Planning*, 23(1), 30-42.

Singh, S. S. B., Kleeman, G., Bergen, P. V. (2012), 'Opportunities to implement GIS in teaching and learning geography: a survey among smart schools in Sabah, Malaysia', *Procedia – Social and Behavioural Sciences* 69, 884-889.

Smith, R. C. (last checked 2012), 'Implementing GIS-based Highway Safety Analysis: Bridging the Gap'. URL. `http://proceedings.esri.com/library/userconf/proc00/professional/papers/pap888/p888.htm`

Sutton, J. C. (1996), 'Role of geographical information systems in regional transportation planning', *Transportation Research Record*, 1518, 25-31.

Sunarto, K. (2011), 'GIS application for land suitability of ginger cultivation'. *10th Annual Conference & Exhibition on Geospatial Information Technology & Applications*, Asia Geospatial Forum.

Solanki, J. & Seetha, U. (2012), 'Geographical information system for power utilities', Global Journal of Computer Science and Technology 12, 53-56.

Steiniger, S. & Hunter, A. J. S. (2012), 'The 2012 free and open source GIS software map – a guide to facilitate research, development, and adoption', *Computers, Environment and Adoption* (in press), 1-15.

Ventura, S. (1995), 'The use of geographic systems in local government', Public Administration Review 55(5), 461-467.

Wheeler, P., Gordon-Brown, L., Peterson, J., Ward, M. (2010), 'Geographical Information Systems in Victorian secondary schools: current constraints and opportunities', *Research in Geographical and Environmental Education*, 19.2, pp.155-170

Xu, L. (2005), 'A decision support model based on GIS for vehicle routing', Proceedings of International Conference on Services Systems and Services Management Shanghai, 2, 1126-1129.

Yap, L. Y. , Tan, G. C. I., Zhu, X., Marissa, C. (2008), 'An assessment of the use of geographical information system (GIS) in teaching geography in Singapore school', Journal *of Geography* 107 (2), 52-60.

Web–based Participatory GIS with data collection on the field – A prototype architecture

by Maria Antonia Brovelli, Marco Minghini and Giorgio Zamboni

Politecnico di Milano – DICA, Italy.
maria.brovelli@polimi.it

Abstract

The rise of Web 2.0 and the current, unprecedented diffusion of mobile devices have laid new foundations for the development of PGIS (Participatory GIS). This study evaluates the possibility of exploiting FOSS (Free and Open Source Software) tools to build up a PGIS prototype providing Web publication of user field-collected data. Besides increasing public awareness and collaboration, user-generated content should also enlarge the knowledge of specific phenomena up to the local level. A prototype architecture was designed and tested in relation to a simple, planning-related case study, i.e., the report of road pavement damages. Open Data Kit suite was used to gather georeferenced multimedia data using mobile device sensors (e.g., the GPS) and to store them into a PostgreSQL database with PostGIS spatial extension. Data was then Web-published using GeoServer. Web access was finally enabled from both traditional desktop-computers and mobile platforms through ad hoc OpenLayers and Leaflet client-side solutions. The architecture provided support for FOSS applicability within the typical PGIS-related tasks, from field survey to data storage, management and dissemination on the Internet. This bottom-up communication paradigm, which exploits real-time, freely available user contributions, can become a potentially precious tool for making decision-processes more democratic, faster and ultimately better.
Keywords: WebGIS, GeoWeb, Web Mapping, FOSS, Mobile, PGIS.

Introduction

The dissemination and sharing of geospatial information on the Web has recently known a tremendous impulse. After the rise of Web 2.0 (O'Reilly 2005), Internet communication paradigm evolved from the traditionally top-down, centrist approach (typical of Web 1.0) to a collaborative, bi-directional model, where users are no longer pure consumers but can also interactively create and share contents (Rinner et al. 2008). The Open Geospatial Consortium (OGC) standards for Web mapping interoperability (Peng & Tsou 2003), together with the non-stop development driven by AJAX (Asynchronous JavaScript and XML) technology, spread this revolution to the field of geographic applications known as Geospatial Web. As a consequence, the terms GeoWeb 2.0 (Maguire 2007) and Web Mapping 2.0 (Haklay et al. 2008) were coined in order to identify the new era of dynamic, interactive tools allowing user participation in managing spatial data. Other concepts introduced after the dawn of GeoWeb 2.0 were neogeography (Turner 2006), i.e., the creation of customized maps through mash-ups of multiple data sources, and VGI (Volunteered Geographic Information), i.e., the idea of humans behaving like sensors able to register geospatial contents. A final, remarkable boost to this new era of Web mapping was given by the tremendous spread of mobile devices, provided with sensors (e.g., cameras and GPS receivers) able to collect georeferenced data and a network connection to immediately publish it on the Web.

The paradigm of Participatory GIS (PGIS) or Public Participation GIS (PPGIS), born in the mid-1990s to define the use of GIS as a tool for promoting citizens' intervention in decision-making processes, found an incredibly fruitful ground in the developments of GeoWeb 2.0. The focus of literature research on PGIS, which is generally related to planning disciplines (Sieber 2006), has therefore dramatically changed in the last two decades. First studies addressed the social and critical role of PGIS (Elwood 2006), paying little attention on the GIS technical aspects. Almost all these applications were desktop-based (e.g., Oliveira et al. 1999) or even paper-based (e.g., Brown & Reed 2000) and simply consisted in getting the public informed about the results of some GIS analysis. Technological developments of Web 2.0 allowed the use of proprietary APIs (Application Programming Interfaces), especially the Google one, for creating interactive Internet applications where users could enter and share customized information (e.g., Bugs et al. 2010, Brown et al. 2012). The use of Open Source Web Mapping tools is still today quite limited due to their higher required skills with respect to APIs (Boroushaki & Malczewski 2010) and only occurred in few literature studies (e.g., Hall et al. 2010). Finally, a growing number of applications (both proprietary and Open Source) have been developed, which allow to Web-access and visualize georeferenced information collected in situ from mobile devices (e.g., Maisonneauve et al. 2010).

Methodology

The purpose of this research is to contribute to the literature on PGIS by using Free and Open Source Software (FOSS) in order to design a participative system architecture able to: a) allow citizens to exploit their mobile devices for collecting geotagged multimedia data (e.g., images, audios and videos) on the field; b) store and manage data into a spatial database and publish it on the Web through OGC standard protocols; and c) set up client interfaces for data visualization on both desktop and mobile devices.

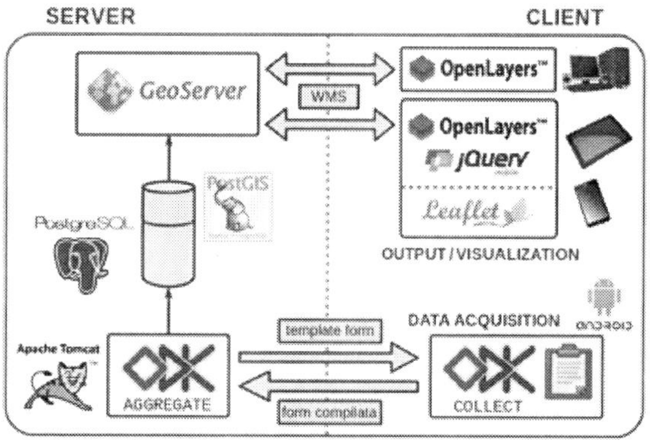

Figure 1: Architecture of the system.

The prototype implemented architecture, divided into the server and the client sides, is shown in Figure 1. Data collection on the field was achieved through the Open Data Kit (ODK) suite (`http://opendatakit.org/`), which provides a modular framework for building specific questionnaires, compile them on the field using Android devices and send them to a PostgreSQL database server (`http://www.postgresql.org/`). Thanks to PostGIS (`http://postgis.net/`), the PostgreSQL spatial extension, attributes can be extended with data geometries and can be read by GeoServer (`http://geoserver.org/`) for Web-publishing as WMS layers. Three viewers finally allow data visualization and querying. The first one was thought for traditional desktop-computers and was built with OpenLayers (`http://openlayers.org/`). The second and the third ones are instead optimized for mobile devices and are based on Leaflet (`http://leafletjs.com/`) and on the OpenLayers mobile version coupled with jQuery Mobile (`http://jquerymobile.com/`).

The applicability of the described FOSS architecture was tested in relation to a simple, local planning issue, i.e., the management of field-collected data concerning reports of road pavement damages. In order to classify these damages, the catalog provided by the Direzione Generale Infrastrutture e Mobilitá (*General Head Office for Infrastructures and Mobility*) of Lombardy Region, Italy, was used (Regione Lombardia 2005). This document sorts pavement deterioration phenomena into three categories: instability influencing pavement surface, instability influencing pavement regularity and cracking. In addition, according to specific qualitative and quantitative considerations, a severity degree equal to low, medium or high can be attributed to the events belonging to each category.

The Build module of ODK suite was first used to design the questionnaire. Users are asked to provide basic information such as the survey date and the address of the pavement damage location. They have then to choose the category corresponding to the damage event and assign it a severity degree equal to low, medium or high. Finally, users are asked to register their current position (using e.g., the device GPS) and to take a picture of the deterioration phenomenon. The ODK Aggregate module, deployed on an Apache Tomcat server (`http://tomcat.apache.org`) backed with a PostgreSQL database, allows then to import and store the questionnaire template created with the Build component. ODK Aggregate allows also to create users and manage their privileges (e.g., to collect data on the field, to view and export data from ODK Aggregate, to upload and delete forms and to manage user permissions). Finally, the ODK client-side Collect module allowed to perform data collection on the field. It consists of an application running on Android devices and synchronized to the Aggregate component. After logging in, users can download the questionnaire template from the ODK Aggregate server, fill the questionnaire on the field and send back the finalized questionnaire to ODK Aggregate. Figure 2 depicts some steps of the questionnaire compilation on an Android smartphone. The survey was addressed to randomly-selected citizens, who were provided with the instructions on how to use the ODK Collect application.

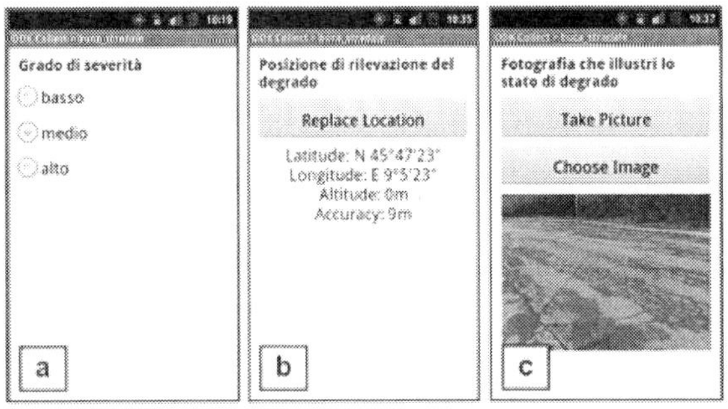

Figure 2: Screenshots of the Android ODK Collect application for reporting pavement damages, showing (a) the choice of the event severity degree, (b) the event position registration using GPS, (c) the event picture taken with the device camera.

User submitted contents, stored into the PostgreSQL database server, become then available into a Web-based graphical application provided by the ODK Aggregate module. Data is organized in a tabular structure, where each submission corresponds to a row and each attribute to a column. Multiple additional attributes (such as the submission identifier, the submission start time and end time) are also added by default from ODK but are not needed for the purpose of the study. For this reason, using also PostGIS for building the geometrical point features, a custom SQL view was created, in order to extrapolate from the previous table only the attributes of interest. Due to its ease of use as well as its certified WMS performance (McKenna 2011), GeoServer was then chosen for publishing the resulting PostGIS vector data as a WMS layer.

Exploiting the OGC Styled Layer Descriptor (SLD) standard (Open Geospatial Consortium 2007), two layer styles were created within GeoServer. The first represents the pavement damage points with flags of different colors according to the event severity degree: green for low degree, yellow for medium degree and red for high degree (see Figures 3 and 4). The second style shows instead the survey positioning precision, which is another attribute recorded by ODK, as the radius of a semi-transparent circle drawn around the points.

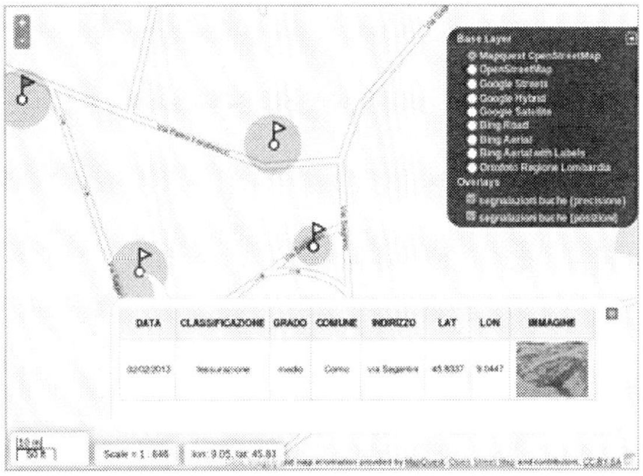

Figure 3: Visualization and querying of pavement damages WMS layer into the client for desktop-computer users.

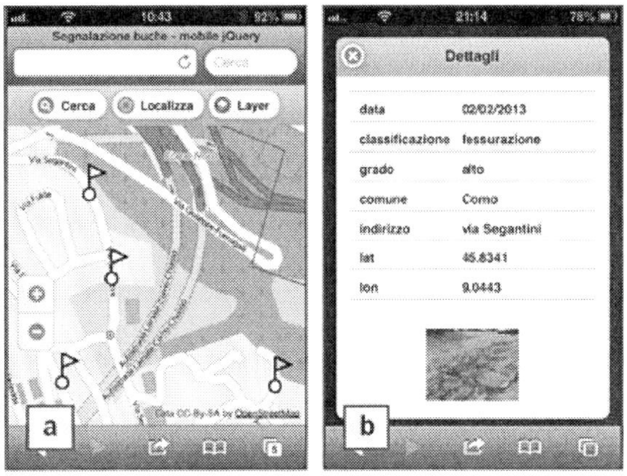

Figure 4: Visualization (a) and querying (b) of pavement damages WMS layer into the client for small-screen mobile devices.

Three viewers were finally created in order to optimize data visualization on desktop-computers, tablets and smartphones. The first viewer exploits OpenLayers JavaScript library to provide data interaction on traditional computers (see Figure 3). The WMS layer representing road pavement damages can be superimposed on different base maps (e.g., OpenStreetMap, Google and the local Lombardy Region orthophoto). When clicking on a point, a WMS GetFeatureInfo request is performed and the answer, i.e., the user field-collected information, is shown inside a popup.

Users accessing the viewer from a mobile device browser are automatically directed to one of two other viewers according to the device type.

For large-screen mobile devices (e.g., tablets), the Leaflet JavaScript library was used to develop a viewer supporting touch-screen commands and providing again WMS GetFeatureInfo response inside a popup. For small-screen devices (e.g., smartphones), a viewer based on the OpenLayers mobile version coupled with the jQuery Mobile Web framework was developed (see Figure 4). Due to the limited screen dimensions, WMS GetFeatureInfo results are now shown into a separate Web page. Some peculiar functionalities of mobile devices, e.g., the representation on the map of the estimated position of the device, are also available within the viewer. The developed mobile viewers were successfully tested on both Apple iOS and Android devices.

Conclusions and future outcomes

This study investigated the development of a Web-based Participatory GIS (PGIS) providing access to field-collected information. Exploiting the idea of participatory sensing (Burke et al. 2006), which conceives human beings as a network of continuously-monitoring sensors, a sample of users was instructed on how to use their mobile devices in order to gather georeferenced data related to road pavement damages. User-generated content was stored and managed into a spatial database and then Web published as a WMS layer. Two main innovations can be pointed out with respect to PGIS literature. The first is represented by system access from mobile devices, which allow to spatially interact with data by means of touch commands. Ad hoc versions were developed for both the large-screen tablets and the small-screen smartphones. Secondly, the implemented architecture was fully developed with Free and Open Source Software (FOSS) tools, which proved to be suitable for the research objective.

In accordance with the fundamental paradigm of PGIS, real-time citizen local knowledge provided a precious means for dealing with decision-making processes in an effective way (Sieber 2006). On the other hand, data collection on the field performed through the ODK Collect application pointed out some limitations. First, road damages positions registered from the user GPS devices proved sometimes to be inaccurate. This was due to both the low performance of the GPS receivers themselves and the non-matching between the survey position (i.e., the position from which the picture is taken) and the position of the photographed damage phenomenon. In order to overcome these issues, as well as to allow users to perform field-data collection even without a GPS-equipped device, an intervention on the Android ODK Collect code is planned, which provides users with an interactive map where they can manually place the point of interest (or refine the GPS-estimated one). Finally, future work should also extend data visualization to the third and the fourth dimension through the use of virtual globes coupled with data spatiotemporal analysis.

References

Boroushaki, S. & Malczewski, J. (2010), 'ParticipatoryGIS.com: a WebGISbased collaborative multicriteria decision analysis', *Journal of the Urban and Regional Information Systems Association* 22 (1), 23–32.

Brown, G. & Reed, P. (2000), 'Validation of a forest values typology for use in national forest planning', *Forest Science* 46 (2), 240–247.

Brown, G., Montag, J. M. & Lyon, K. (2012), 'Public Participation GIS: A method for identifying ecosystem services', *Society & Natural Resources* 25 (7), 633–651.

Bugs, G., Granell, C., Fonts, O., Huerta, J. & Painho, M. (2010), 'An assessment of Public Participation GIS and Web 2.0 technologies in urban planning practice in Canela, Brazil', *Cities: The International Journal of Urban Policy and Planning* 27 (3), 172–181.

Burke, J., Estrin, D., Hansen, M., Parker, N., Ramanathan, S., Reddy, S. & Srivastava, M. B. (2006), Participatory Sensing, *in* 'Workshop on World-Sensor-Web (WSW'06): Mobile Device Centric Sensor Networks and Applications (Boulder, October 2006)', Boulder, CO, 117–134.

Elwood, S. (2006), 'Critical Issues in Participatory GIS: Deconstructions, Reconstructions, and New Research Directions', *Transactions in GIS* 10 (5), 693–708.

Haklay, M., Singleton, A. & Parker, C. (2008), 'Web Mapping 2.0: The Neogeography of the GeoWeb', *Geography Compass* 2, 2011–2039.

Hall, G. B., Chipeniuk, R., Feick, R. D., Leahy, M. G. & Deparday, V. (2010), 'Community-based production of geographic information using open source software and Web 2.0', *International Journal of Geographical Information Science* 24 (5), 761–781.

Maguire, D. J. (2007), GeoWeb 2.0 and Volunteered GI, *in* 'Workshop on Volunteered Geographic Information', University of California, Santa Barbara, 104–106.

Maisonneauve, N., Stevens, M. & Ochab, B. (2010), 'Participatory noise pollution monitoring using mobile phones', *Information Polity* 15 (1), 51–71.

McKenna, J. (2011), 'WMS Performance shootout 2011', Denver, FOSS4G 2011. URL: `http://www.slideshare.net/gatewaygeomatics.com/wms-performance-shootout-2011` (Last checked May 2013).

Oliveira, M. P. G., Medeiros, E. B. & Davis, C. A. Jr. (1999), Planning the acoustic urban environment: a GIS-centered approach, *in* 'Proceedings of the 7th International Symposium on Advances in Geographic Information Systems (Kansas City, 2-6 November 1999)', Kansas City, USA, 128–133.

Open Geospatial Consortium (2007), *Styled Layer Descriptor profile of the Web Map Service Implementation Specification*, Version 1.1.0 (revision 4), OGC 05-078r4 . URL: `http://portal.opengeospatial.org/files/?artifact_id=22364` (Last checked May 2013).

O'Reilly, T. (2005), 'What is Web 2.0: Design patterns and business models for the next generation of software'. URL: `http://oreilly.com/pub/a/web2/archive/what-is-web-20.html` (Last checked May 2013).

Peng, Z. R. & Tsou, M. H. (2003), *Internet GIS*, Hoboken, NJ, John Wiley & Sons.

Regione Lombardia – Direzione Generale Infrastrutture e Mobilitá (2005), *Catalogo dei dissesti e delle pavimentazioni stradali, ALLEGATO B*, d.g.r. n. 1790/2006, BURL n. 8 S.S. Del 21.2.2006.

Rinner, C., Ke§ler, C. & Andrulis, S. (2008), 'The use of Web 2.0 concepts to support deliberation in spatial decision-making', *Computers, Environment and Urban Systems* 32, 386–395.

Sieber, R. (2006), 'Public Participation Geographic Information Systems: A Literature Review and Framework', *Annals of the American Association of Geography* 96 (3), 491–507.

Turner, A. (2006), *Introduction to Neogeography*, Sebastopol, CA, O'Really Media Short Cuts.

Taarifa

Improving Public Service Provision in the Developing World Through a Crowd-sourced Location Based Reporting Application

by Mark Iliffe[1]*, Giuseppe Sollazzo*[2]*, Jeremy Morley*[1]*, and Robert Houghton*[1]

*1: University of Nottingham (UK); 2: St. George's University of London (UK).*psxmi@nottingham.ac.uk

Abstract

Public service provision in the developing world is challenged by a lack of coherence and consistency in the amount of resources local authorities have in their endowment. Especially where non-planned urban settlements (e.g. slums) are present, the frequent and constant change of the urban environment poses big challenges to the effective delivery of services. In this paper we report on our experiences with Taarifa: a location-based application built through community development that allows community reporting and managing of local issues.

Keywords: Location-based application, community development, crowd-sourcing.

Introduction

The availability of geographic data in the developing world is improving with the advent of community mapping projects like Map Kibera [1, 2] and through organisations like Humanitarian Open Street Map Team (H.O.T). Previously a large barrier for NGOs, governments and business in providing services in developing world, the lack of governmental and non-governmental data is becoming an issue of the past and with this barrier rapidly dissolving further questions are arising, such as: now we have the data, what do we do next?

The Taarifa project aims to address this question, with respect to the monitoring of public service provision. Taarifa as a software platform allows for the community reporting of problems, from health to waste issues, through a mobile phone interface using SMS or a HTML5 client. Once reports are collected they are entered into a workflow allowing those in charge of providing services to monitor, triage and act upon reports.

Taarifa is currently unique in this field from its initial design, inception and deployment. It was originally conceived at the Random Hacks of Kindness (RHOK) London Water Hackathon. A Hackathon is an organised meeting of developers who team up to code on a specific topic or to address a specific problem. Hackathon are very popular in the developers community. Both private and public organisations often set up hackathons to get some fresh hands working on certain issues they are facing. During the RHOK hackathon a group of core developers 'hacked' a solution in 48 hours. After continued development and design Taarifa was first deployed in cooperation with the Ugandan Ministry of Local Government in March 2012, followed by a deployment with the Zimbabwean Government in April 2012 facilitated by the World Bank.

In this paper we present a narrative and case study of the Taarifa project from its inception, design, refinement and deployment. We also discuss the future directions Taarifa might take both as a community software project and as an organisation. The global aim is to facilitate a discussion on how crowd-sourced geospatial data and open source platforms can combine to improve public and private service delivery in developing nations.

Related Work

Most related work considering the emergent phenomena of crisis mapping are case studies of specific crises, the Haiti Earthquake [3], terrorist attacks [4], methodologies for crisis situation triage [5]. Though, in context not all methodologies of crisis relief are wholly focused on external response as [6] demonstrates. These instances of citizens generating reports fall under the banner of crowdsourcing. Here [7] specifically looks at providing situational awareness; how a visual 'group map' of all the reports is useful as errors can be made. The crowd sourcing of information has inherent dangers of trusting the information supplied, as [8] demonstrates with respect to potential crisis situations.

Currently there are two themes missing from the literature; A study of how the reports are being used to aid decisions and an understanding of areas in a constant state of crisis. These areas, like slums and informal developments do not have an event like an earthquake or a tsunami to illustrate the plight. This is combined by [9] review of the Map Kibera project.

Almost anecdotal evidence [10, 11] exists to how the crowd sourced data is used, but nothing to the experiences of using it, within the field of crisis mapping. However, numerous papers cover the experiences of the crowd and their validation in citizen science, like ecology [12]. A possible output and definite gap in the research would be an ethnographic study of when a crisis occurs, observing and reporting the 'value chain' and how the data is used. A noteworthy omission from this section of the review is ICT4D field of study. While this is a rapidly developing field the research seems to be based more on the social science fringe compared to the more software/algorithm development side, though the gap is being bridged.

The effect of social media, indicates that on a social level it aids the transition to recovery through blogs [13, 14] and as a general community platform to generate maps [10]

Taarifa's inception at the RHOK Hackathon

The London Random Hacks of Kindness Hackathon (RHoK) occurred on Friday 21st of October, 2011, lasting for 48 hours in the facilities of University College London. Randomly assembled groups of coders with interests on humanitarian subjects and matter experts joined forces to work together with the aim of producing technology demonstrators and designs to solve problems related to water.

Why A Hackathon?

The underlying idea of any hackathon, and in particular of the RHOK Hackathon, is that by co-locating intelligent, innovative and driven computer software developers and field experts, facilitating a fast-paced production session lasting no more than a couple of days, can lead to worthwhile innovation. One condition is that all software produced has to be open source, and this naturally offers opportunity of further development by different teams in the future. There are many kinds of hackathons: from those destined to the very young, to corporate-run hackathons in which participants are selected and paid a daily fee. Many question have arisen about the effectiveness of hackathons in producing something really useful, especially generalist hacking events. However, we argue that dedicated hackathons like the RHOK are a practical and effective way of developing products, mostly due to their very narrow target. They can be seen as "sprints" sessions as formalised by the Agile project management theory [15].

A story telling of the event

The event started with an introduction by the organisers framing the exercise with problem statements and presentations by experts wanting to solve different problems. Some of these people spoke in person at the event, while others via teleconference. The problems presented all came form the RHoK website and ranged from water trading platforms to public service infrastructures and community mapping.

Figure 1: London WaterHackathon 2011

Taarifa started to take shape of a platform to support citizen interaction around public services for lesser developed countries.

A team was assembled and collectively started to set up whiteboards and tools to construct the intended processes and a design specification for workflows through the potential system. The discussion focussed on how triaging of reports would work. At the time Taarifa's intended customer was meant to come from a ministry level, potentially working around sanitation or waste issues.

Aiming for rapid development of the platform we decided to fork the Ushahidi platform (Okolloh 2009). Ushahidi is a platform/CMS designed for the crowdsourced reporting of issues, its inception was due to the Kenyan election crisis of 2008, since has been used to report conflicts like civil wars but also the recent 'occupy' movements. Technologically it's foundations are built in PHP using the Kohana framework. Here we hit our first issue; none of our developers had worked with Kohana before. We split into two groups, one figuring out Kohana, the other designing workflow.

It became apparent that experience in the right tools was needed with some developers wishing to contribute. However, some were not able to operate at the level as some developers, which is typical of hackathons where a large team works together. As this progressed, these developers filled other roles, testing and aiding the main thrust instead of contributing code. However, their contribution was as valuable in real terms as the code generated. Installing Ushahidi also proved problematic. Issues with mod_rewrite and other PHP extensions were experienced, but were eventually resolved. These potentially could have been avoided through enhanced documentation. Equipment at the hackathon was problematic: the team didn't have access to hosted server, hence one of the developers' personal servers was used.

Once the workflow was sketched out we presented back to the other team of developers. They had conducted a study into the Ushahidi plugin ecosystem. Collectively we integrated the 'actionable' and 'simple groups' plugins. Actionable was adapted to 'action' reports, and place them in the triage system. Simple groups was used to curate a team of 'fixers'. Fixers was used generically as the people fixing the reported problems, however the dynamic of how this would be fully implemented wasn't considered at this stage.

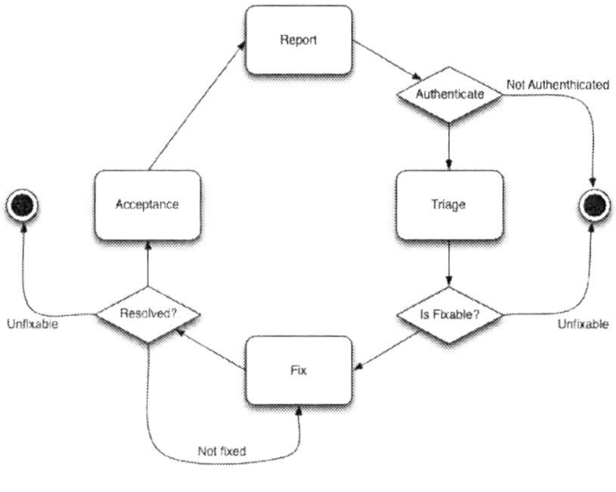

Figure 2: Workflow Of Reports Through The Taarifa System

Inheriting the interface from Ushahidi meant the load on the developers was focused towards the back-end with tasks and problems being received and triaged. The workflow started to come together, based on the idea of community reports being verified, then put on triage, assigned to a team of fixers and finally reaching conclusion or, if not appropriate, directed to dispute resolution. The visual user interface, organised in tabs to accommodate these functions, was integrated into the system. As reports were able to be triaged we focused on expanding the reporting mechanism. Ushahidi supports reporting through a web-based form, twitter and through its mobile applications (iOS, Android, Java and Windows Mobile 6). It can interface with SMS gateways like FrontlineSMS. The team intended to use SMS due to the ubiquitous nature of feature phones in Africa that realistically can only use SMS as a form of reporting. Using SMS presented problems of geolocating the messages.

The OGC [16] standard on GeoSMS was unfortunately unavailable at the time, it is possible for the mobile phone networks to triangulate the position of the sender and supply a latitude and longitude however this isn't practical over a 48 hour hackathon notwithstanding the ethical and privacy concerns.

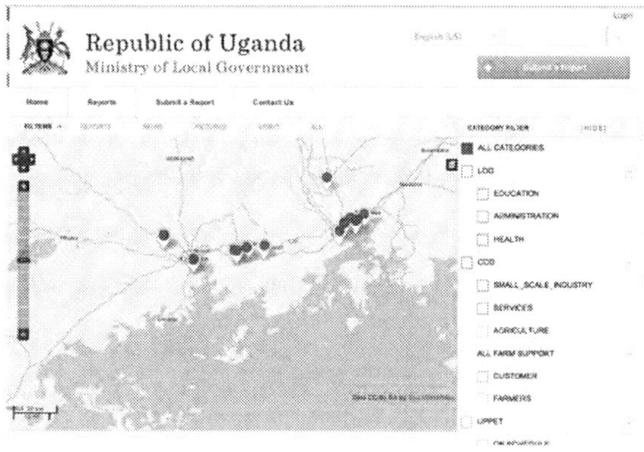

Figure 3: Taarifa Interface

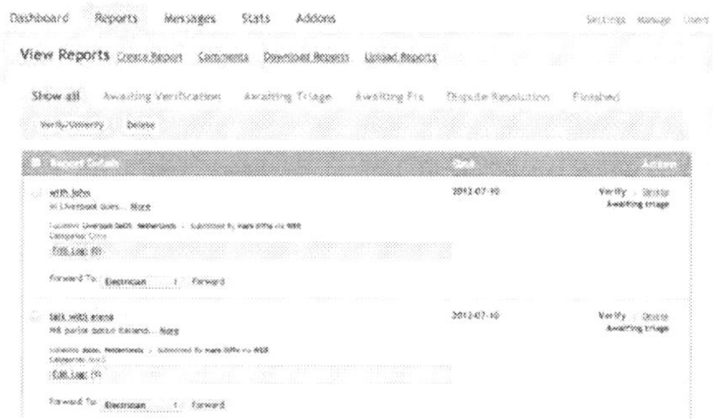

Figure 4: Taarifa Workflow Management

In response to this issue a $100m^2$ grid was created under a custom coordinate reference system having a 10 digit reference for each grid square. Then a reporter with a SMS capable phone would input the number with a hash (#) then found through a regular expression in the submitted message. Obviously questions remain when implementing this on a large scale namely ensuring local people know what their code is and creating a reference system that conforms to the human geography not just the physical. This was accomplished on the first day of the hackathon, however we worked through the night, resting for four hours. A global team of friends carried on completing an SMS gateway. What remained at the hackathon around bug-fixing, tidying up code starting documentation and choosing a name. One team member searched the word "Reporting" into a translation software for the Swahili language. 'Taarifa' was the result.

Discussion about the results of the hackathon

The Taarifa group at the hackathon was fortunate to be successful and be voted winners of the London hackathon. It was decided that the project was an interesting effort to address real problem, and should be kept alive. The team had worked and synchronised well. Administratively an online mailing list was created, communication through instant messaging and logos and branding. The team assessed that the integration of the mobile applications was key, though development was dispersed over the Android, iOS and Windows Mobile platforms. A decision was made to focus on a web-based HTML5 application. Using the offline functionality of HTML5 and CSS3 a mobile application was quickly prototyped.

Deployment: Uganda

The Africa Urban and Water (AFTUW) sector of the World Bank approached the Taarifa project about a pilot with the Ugandan Ministry of Local Government (MoLG). The ministry wished to monitor local government projects based around improving community cohesion, public services and enterprise. A pilot in four districts was decided upon as part of the "Improving Systems for the Urban Poor" of AFTUW supporting two ministry led programs Community Driven Development (CDD) and Local Government Management and Service Delivery (LGMSD).

CDD is a match funding program where community members form groups around themes of entrepreneurship, farming and education. Funding is given in ratios of 2-5:1 and are aimed at directly improving the development of communities. LGMSD is a government program aimed at building capacity within government. For instance the building of council facilities and schools.

Figure 5: Reporting.

Figure 5 shows the reporting of a local government building in construction by civil servants. Traditionally the system of CDD and LGMSD was paper based. These forms were then posted from the areas to central government in Kampala. This drawbacks, first was the postal service of Uganda with idiosyncratic delivery. Second was the load placed upon the reporters. The complex nature of the questions posed by CDD and LGMSD posed difficulty to civil servants, who may not have the appropriate equipment to submit forms.

Civil servants reporting to these programs were selected for training, with AFTUW supplying Android based Hauwei 'Gaga' mobile phones - under $100 - as the hardware platform. Initially the custom forms of Taarifa worked well, with the participants able to submit information. However, when venturing into more remote districts the functionality of Taarifa inhibited reporting, specifically offline forms. While reporting offline is possible, it isn't possible to change forms without connectivity. This requirement since entered the Github repository; the platform used for the project management of the Taarifa platform.

Improvements were identified by the pilot and were fed back into the Taarifa community. However the pilot in the four districts was deemed successful by MoLG and AFTUW, consequently the platform was rolled out to the 111 districts of Uganda. Currently the system is directly administered by MoLG, however they are activity seeking a devolution of control to the local districts. This in itself will be a large undertaking and one which potentially requires more structure than the Taarifa project in its current form can provide.

Evaluation and comparison of community software contribution

Given the way Taarifa has been developed, a question emerges as to how to best evaluate the community effort behind this development. To address this question, we need to inspect the contributions made to the open source project and try and make sense of these. It is important to remember that Taarifa was born as a fork of the Ushahidi project, and originally called Taarifa Web. This is the version in use in most deployments. A newer version, called Djangorifa, is a complete rewrite from scratch of Taarifa, using the Django framework in order to achieve a lightweight distribution. The rewrite also aims at freeing Taarifa from functionalities and features which were of use only to the goals for which Ushahidi was developed. In this section we will present a comparison of these three projects based on code contribution statistics.

Data collection methodology

All of the three projects, Ushahidi, Taarifa Web, and Djangorifa, are community open source project. They all use the same set of online tools and services in order to allow collaborationg: GitHub, an online code repository based on the git protocol. The protocol allows us to collect data which can be used to extract statistics. Using the `git log` and `git shortlog` commands, we were able to identify the number of commits, files changed, lines insert and lines deleted for each contributor. From this data we extracted information like the number of contributors, the lines committed, the number of commits, and calculated average contribution, standard deviation, and the relative quintiles. This methodology has its limitations, intrinsic to the type of projects we are discussing. For example, it must be remembered that the Taarifa Web project comes from a fork of Ushahidi, hereby sharing much of the codebase; in this analysis, we focus on the code contributed after the fork because we can then see how much code has been genuinely added. Also, we use quintiles analysis because we are talking about different amounts of code contributed: given that Taarifa embeds much of the Ushahidi codebase, having a quintiles analysis on the differential contribution allows us to compare the community involvement and interactions. The Djangorifa project was born as a single-person effort and is a relatively smaller project. Still, we add it for completenes and comparison. Although we cannot claim this is a complete analysis, it is helpful to identify trends and customs in community contributions. Data are summarised in Table 1.

Table 1: Summary of project statistics.

Param	Ushahidi	Taarifa Web	Djangorifa
contributors	71	12 (83)	5
commits	4017	667 (4684)	55
branches	8	6	4
lines of code	>220,000	>40,000 (260,000)	>35,000
average commits	56.58	7.53	11
std. dev. commits	154.66	5.06	15.35
average LOC inserted	13401.64	2935.92	5927.4
std. dev. LOC inserted	37496.32	9966.27	10553.19
issues (open)	240	21	11
issues (closed)	779	29	4

Ushahidi

A total of 71 developers have contributed to this project, which is a medium-large software project, also highlighted by the high number of lines of code (in excess of 240,000) and commits (over 4,000). With such a great number of contributors, we can see very large variations in their contributions. Despite the largest single contributor having committed almost 800 times with over 40,000 lines of code (LOC), the average is about 57 commits per contributor, with a very high standard deviation of 154.66; the same applies to the LOC, where we see an average of 13401.64 and a standard deviation of 37496.32. This shows that the contributions have been varied across the community, although we have a strong core of developers who have contributed a huge amount of LOC. An interesting analysis can be that considering the quintiles. For example, where commits are concerned, figure 6 shows how many contributors fall in each quintile. The quintile range is defined in the caption of the figure. It is evident that there is a relatively high amount of people only contributing a small number of times to the projects (20th percentile). The number then drops for the 40th percentile and slowly increases up to the 100th percentile. This is consistent with a community where there is a core of steady contributors collaborating with a galaxy of less involved contributors. However, if we apply the same analysis to the number of LOC contributed (see 7), we find that there is a certain balance in the number of contributors within each quintile. This suggests that the amount of effort each contributor is able to provide the community is similar, although sometimes spread over multiple commits. There is, however, a medium correlation (correlation coefficient of 0.49), between the commits and the LOC samples, which is consistent with an ongoing effort.

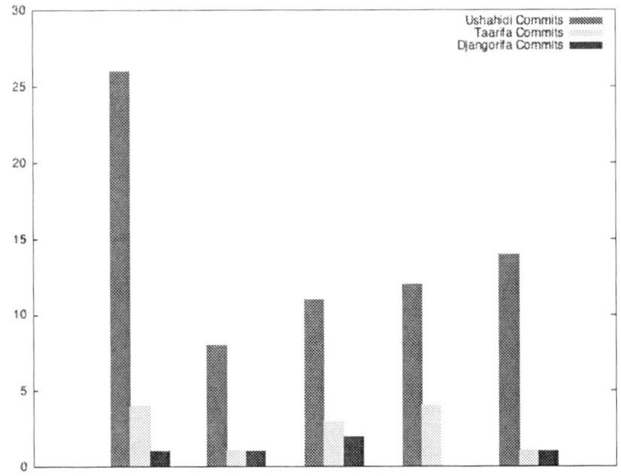

Figure 6: Quintile cardinality for commits; Ushahidi [1,2,5,34,777], Taarifa Web [3,5.8,8.2,12,18], Djangorifa [1.8,5,7,13.2,38]

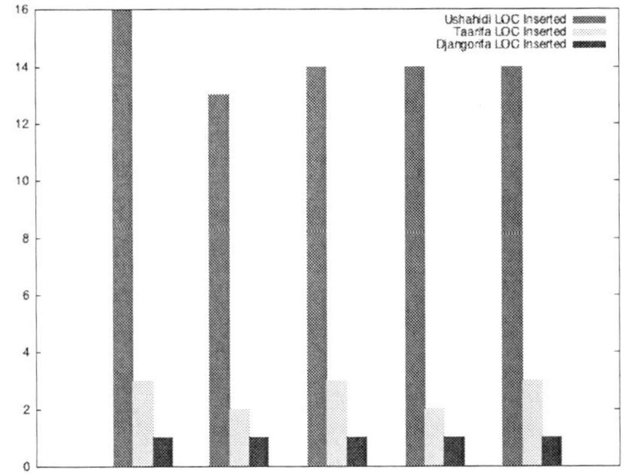

Figure 7: Quintile cardinality for number of lines of code inserted; Ushahidi [6,26.4,226.6,9706.4,229389], Taarifa Web [4,28,190.6,481.2,36098], Djangorifa [50.2,132.8,2087.6,8839.8,24427]

Taarifa Web

The Taarifa community is much smaller than the Ushahidi community, and the total amount of LOC is bigger only because Taarifa Web was originally forked out from Ushahidi. In fact, Taarifa adds about 40,000 LOC on top of the Ushahidi code base, coming to a total of 260,000. The number of contributors who worked exclusively to Taarifa is 12, coming to a total of 667 commits. Clearly, this implies that in its shorter life each contributor had fewer opportunities to commit code: the average per-contribute is 7.53 with a standard deviation of 5.06. This suggests more uniformity in the way Taarifa Web was developed and is in fact consistent with its inception at a hackathon: there is a relatively high number of people doing commits, rather than the massive differences we saw for Ushahidi. The story is a bit different when we analyse the LOC contribution. With an average of 2935.92 and a standard deviation of 9966, we can still see wild differences in the levels of contribution. This finding can be explained in a very simple way: Taarifa is a community effort were there is a core groups of developers, and most importantly a lead developer. The lead developer is most definitely an outlier, having contributed a whopping 36098 LOC. In this case, there is a clear concentration of many LOC in just a few commits, typical of hackathon developments. This is confirmed by the low correlation (0.28) existing between the commits and LOC samples. Visual confirmations of this can be found in figure 6 and 7.

Djangorifa

As stated before, we only add this for completeness. Djangorifa is a novel implementation of Taarifa based on a new framework and it is pretty much an ongoing effort. The data we see here are consistent with an ongoing personal developmental effort, especially an extremely low correlation coefficient (0.05) between the commits sample and the LOC.

Discussion

The most important lesson we gather by analysing these data is that we can distinguish an ongoing effort, spread over a large community, that of Ushahidi, from an development like that of Taarifa, which had an initial large contribution at a hackathon, upon which a small community then developed. Also, we are able to identify where there is a core of developers, a lead developer, or a single developer, as in the case of Djangorifa. Further analysis might lead us to identify trends in Taarifa, seeing when and where contributions are more likely to give a lasting effect on the community.

Summary and Future

This paper has discussed how the Taarifa project was started and how it was used in Uganda. Issues identified with the deployments include problems with civil infrastructure and communications in the country. While it is realistic to adapt the Taarifa platform to be resilient with regard to poor connectivity it will presumably be an issue which will need to be addressed in future iterations.

How can the Taarifa platform deal with an environment with no connectivity? We are currently assessing and piloting the use of Taarifa in Tanzania and in the United Kingdom, the adaption in these environments, with differing infrastructure and political will, remain outstanding. Taarifa as a group is currently looking towards formalising as an organisation. As an open source movement it can go so far, however as a loose collection of interested humanitarians the project can only go so far. For example documentation is an area which is in need of improvement, not just in requirements but user guides and manuals of use. Unfortunately "the code is the documentation" isn't an approach that the Taarifa project wishes to take.

As an organisation, formal structures and roles can aid in shaping the project. Requirements gathered in collaboration with users of the platform at the ministerial and local level could be investigated with the funding to explore those opportunities. Taarifa is an open source platform and project and is free to download and use. However the time and equipment spent on the project is costly. The community, however, has proved successful in providing enough motivation to the members to keep working on the project.

References

[1] E Hagen. Putting Nairobi's Slums on the Map. Development Outreach, 12(1):41–43, 2009.

[2] M Iliffe. When Government 2.0 Doesn't Exist : Mapping Services In The Developing World. In Proceedings of the 2011 AGI GeoCommunications Conference, 2011.

[3] Matthew Zook, Mark Graham, Taylor Shelton, and Sean Gorman. Volunteered geographic information and crowdsourcing disaster relief: a case study of the Haitian earthquake. World Medical & Health Policy, 2(2):7–33, 2012.

[4] Nathan Schurr, J Marecki, and M Tambe. The future of disaster response: Humans working with multiagent teams using DEFACTO. In Proceedings of the AIII Spring Symposium, 2005.

[5] K Lorincz, D J Malan, T R F Fulford-Jones, A Nawoj, A Clavel, V Shnayder, G Mainland, M Welsh, and S Moulton. Sensor networks for emergency response: Challenges and opportunities. Pervasive Computing, IEEE, 3(4):16–23, 2005.

[6] A Heijmans and L Victoria. Citizenry-Based & Development-Oriented Disaster Response. Centre for Disaster Preparedness and Citizens' Disaster Response Centre, 2001.

[7] L Gunawan, A. H. J. Oomes, M Neerincx, WP Brinkman, and H Alers. Collaborative situational mapping during emergency response. Interacting with Computers, 2009.

[8] P Meier. Do "Liberation Technologies" Change The Balance Of Power Between Repressive States And Civil Society? PhD thesis, Tufts University, 2011.

[9] Erica Hagen. Mapping Change: Community Information Empowerment in Kibera (Innovations Case Narrative: Map Kibera). Innovations: Technology, Governance, Globalization, 6(1):69–94, January 2011.

[10] Rebecca Goolsby. Social media as crisis platform: The future of community mapscrisis maps. ACM Transactions on Intelligent Systems and Technology (TIST), 1(1):7, October 2010.

[11] E Christophe, J Inglada, and J Maudlin. Crowd-sourcing satellite image analysis. In Geoscience and Remote Sensing Symposium (IGARSS), 2010 IEEE International, pages 1430–1433. IEEE, 2010.

[12] Leska S Fore, Kit Paulsen, and Kate O'Laughlin. Assessing the performance of volunteers in monitoring streams. Freshwater Biology, 46(1):109–123, 2001.

[13] B Al-Ani, G Mark, and B Semaan. Blogging in a region of conSSict: Supporting transition to recovery. In Proceedings of the 28th international conference on Human factors in computing systems, pages 1069–1078. ACM, 2010.

[14] T. N. Smyth, J Etherton, and M. L. Best. Moses: Exploring new ground in media and post-conflict reconciliation. In: Proceedings CHI 2010, 2010.

[15] Robert Cecil Martin. Agile software development: principles, patterns, and practices. Prentice Hall PTR, 2003.

[16] K. Mei Chen and C. Reed. OGC: Open GeoSMS Standard-Core. Technical report, Open Geospatial Consortium, 2012.

OSM–GB

Using Open Source Geospatial Tools to Create OSM Web Services for Great Britain

by Amir Pourabdollah

University of Nottingham, United Kingdom.
amir.pourabdollah@nottingham.ac.uk

Abstract

A use case of integrating a variety of open-source geospatial tools is presented in this paper to process and openly redeliver open data in open standards. Through a software engineering approach, we have focused on the potential usability of OpenStreetMap in authoritative and professional contexts in Great Britain. Our system comprises open source components from OSGeo projects, the Open Street Map (OSM) community and proprietary components. We present how the open data flows among those components and is delivered to the Web with open standards. Apart from the cost issues, utilizing the open-source tools has offered some distinct advantages compared to the proprietary alternatives, if any was available. At the same time, some technical limitations of utilizing current open-source tools are described. Finally a case study is shown for the usability of the developed solution.

Keywords: OpenStreetMap, open source, open data, open standards.

Introduction

Since its inception in 2004, OpenStreetMap (OSM) has become the main free source of digital maps made by the crowd. Although OSM is rapidly growing in both contents and contributors, the belief that it is made by amateurs is perceived to limit trust in the value of this free data source within the traditional GIS community. The quality aspects of OSM have been investigated by different researchers and with different tools (Girres and Touya 2010; Haklay 2010; Zielstra and Zipf 2010; Mooney and Corcoran 2011). We contend that to encourage uptake of data, not only must the OSM community produce better maps but the authoritative and professional use cases of OSM shall be facilitated.

Because of the volunteer nature of the community, many open-source geospatial tools have been developed around OSM in the recent years, ranging from data handling and data analysis to cartography and presentation. There are a number of core open-source tools that are used by the OSM developers, e.g. Mapnik (Pavlenko 2011) for rendering, while some other open-source tools have been developed for users and contributors e. g. JOSM (JOSM 2012) and the OSM plug-in for Quantum GIS (QuantumGIS n.d.).

Although those open-source tools generally fit the purposes of core OSM users and contributors, they may not necessarily fit for the purposes of professional map consumers, authoritative users and national agencies. If OSM is not effectively usable by this group, the gap between professional and amateur map producers/consumers may never be filled and a sense of trust in OSM may never happen among those users. On the other hand, if OSM can be used effectively by authorities, there will be a big chance that those users become active contributors, leading to even more usability and reliability.

Having observed this research gap, this paper presents a system with a strong reliance on open source geospatial components that can fit the British authoritative user's requirements. The solution is developed within the framework of a project called OSM–GB (OSMGB 2012).

In the rest of this paper, a software engineering approach is taken to analyze the specific requirements of authoritative users in Britain. Based on those requirements, the conceptual and detailed designs of our system are presented. The solution developed also benefits from open standards and open data initiatives as will be explained later.

Users' Requirements

In the Great Britain context, the following requirements were collected from authoritative users in the first stages of the project, particularly through steering group meetings and direct contacts. The requirements are a combination of general and GB-specific functionalities, as will be explained in this section.

It is however noticeable that gathering the users' requirements is done within the open-source/open-data contexts and implications, i. e. the users are generally willing to use the free software, data and standards as opposed to the licensed products (for many reasons including the known cost and updatedness issues) but they are looking for the best possible solution that fits their specific purposes. Otherwise, the designed system may not address the requirements from the users that wish to benefit from the contractual or the quality-guaranteed services offered by the available licensed products, e. g. the licensed map data from Ordnance Survey.

Coordinate Reference System

In the context of Great Britain, one of the main concerns in using a map in authorities is the availability of the map in the British National Grid (BNG or EPSG:27700) coordinate system. The OpenStreetMap website shows the map in Google Mercator (also called EPSG:900913), as is common in other public maps (e. g. Google and Bing maps). The OpenStreetMap tile service which can be used as the background in desktop GIS applications, is also only in Google Mercator. The export facility of OpenStreetMap website, as well as the OSM "planet files" and replications use OSM-XML format, in which the coordinates are encoded in Lat/Lon (EPSG:4326). There is other no free public service, at the time of this paper, which can deliver OSM data in the BNG coordinate system.

Multiplicity of Layers

Unlike public users, the authorities are specialized and need to have more flexibility in selecting the map features they require. They need the map to be thematically separated in layers, so the users do their job by superimposing the required layers. For vector maps, thematic filtering of the features is commonly feasible in GIS applications, but for raster maps such filtering is usually impossible. Although the features of OpenStreetMap are tagged and the tags can be used to separate them into multiple layers, the OSM website renders them all into a flat map. At the time of this paper, there is no free public service offering multi-layered raster map for OSM. It is however possible to use the filtering functionality of the OSM API, subject to knowing the required tag combination for the specified thematic map, which produces raw text output.

Raster and Vector Maps

Raster maps are necessary but not enough if detailed map analysis is required in the authoritative use cases. Querying and filtering by geometry or by attributes, applying routing algorithms, customized rendering and labeling are examples of the functionalities that can only be achieved in vector maps. OSM-XML is not a known format in our target authorities. At the time of this paper, there is no other free public web service of vector OSM data for the GB area, particularly in BNG.

OGC Compliance

The standards established by Open Geospatial Consortium are the most widely accepted standards for map delivery via the Web. The availability of OGC compliant Web Services is a major advantage, if not a necessity, for a map service to be used by the authorities. The main OGC standards for map delivery are WMS (Web Map Service), WMTS (Web Map Tile Service), WMSC (Web Map Service, Cached) and WFS (Web Features Service). The first three standards are for raster map delivery and the last makes the vector text based output in a number of established OGC standards including GML (Geographical Markup Language) and KML (Keyhole Markup Language).

WMTS is currently run by the OpenStreetMap and some other mirrors in limited CRSs. Besides that, at the time of this paper, the only available world-wide WMS service is run by the University of Heidelberg called World OSM-WMS (GIScience 2010), which is single-layer and is updated weekly. No other WFS based on OpenStreetMap has been found for the GB area.

Consistent Access to National Open Maps

In Britain, Ordnance Survey provides a set of open data (OS 2013) in different formats and resolutions. The authorities may have full access to the Ordnance Survey's open and non-open maps, however if they are willing to work with OSM they prefer to have an integrated access to those national maps in the same platform, particularly for superimposing the maps from different sources. The project's focus has been on open data sets, thus integrating some parts of the OS Open Data with OSM and delivering them in a single platform can enhance the usability of the final product. Among the OS Open Data products, VectorMap District (VMD) is the most comparable to OSM and so was selected to be delivered via the OGC Web Services (WFS in this case) together with OSM. At the time of this paper, no free WFS access to the VMD is found (although the University of Edinburgh's Digimap OpenStream service offers WMS access to OS Open Data, free for UK academic use).

Frequent Updates

The other main concern of the authoritative users is the frequency of updating the map information. A main advantage of crowd-sourced maps like OSM is the near-instant reflection of changes, in contrast to the long bureaucratic process of map updates in authority-sourced maps. After a user adds or changes data in OSM, the system is able to update its output in a very short time difference, perhaps in a few minutes. Any OSM redelivery service should therefore also update its content very frequently.

Quality Check and Fix

As a value-added feature, the OSM quality can also be checked in a defined geo-processing algorithm. Although the project has an extensive subsystem for data quality check and fix, this paper is not going to cover this subsystem in details. Particularly we do not cover the details of the user's requirements in this subsystem. It is however, an integrated part of the system design and the interactions of this subsystem with the rest of components are briefly mentioned.

Briefly, this functionality can benefit both the authoritative and public OSM contributors. The authoritative users can then have a gauge of OSM reliability, and the OSM contributors (including the authoritative users) can have guides to correct the map errors. The main requirement is to have a "rule-based" checking and fixing algorithm. It is not desired to have a closed hardcoded set of rules; instead, they shall be defined as openly and as flexibly as possible. The results of the quality checks and fixes are also desired to be delivered by some integrated and consistent Web Services.

It should also be noted that we are not claiming OSM to be a direct alternative to national mapping data such as that from the Ordnance Survey. The different sources have different characteristics: prime amongst these is the more precise (but closed) data capture & production specifications of the Ordnance Survey which guarantee certain levels of data quality in their data, particularly for example in completeness of coverage.

Summary

According to the items above, the current OSM website, its replications and redelivery mirrors are not likely to be effectively used by the authoritative map users in GB. A system is required that can deliver the OSM data in BNG reference system, in vector and in thematic raster layers, by OGC-compliant means and with frequent updates. Finally it is desired to have consistent access to the OS Open Data and to have a quality checking subsystem.

The Conceptual Design

In order to fulfill the above requirements, a set of software components, including data sources, tools, databases and standards shall put together. Use of open-source tools has been a preference but not an absolute necessity in this project.

At the conceptual level of system design, Use Cases can show the interaction of the system with the external entities. In addition to the users who are the main actors, an automatic update scheduler can be considered as another actor. The OSM data source and its regular updates are the external data to be fetched regularly by the system. Our Use Case diagram is illustrated in Figure 1.

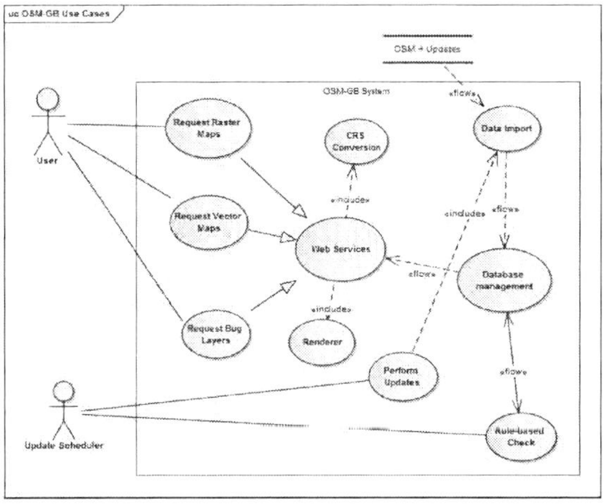

Figure 1: Use-case diagram for the conceptual level of design.

Users request the maps in raster or vector formats through the Web Services. In the case that they work with their preferred choice of GIS application, the application requests the information on their behalf. Moreover they may need to request for the detected "buggy" features which are yet another form of Web Service interaction. The Web Service provider takes its data from the central database management system, and also includes a CRS conversion activity. In addition, it needs to include a renderer component that can make cartographical outputs for the incoming raster map requests.

The "Update Scheduler" actor frequently triggers an update signal. This signal primarily causes a new data import activity that updates the database with the new OSM replication; secondly it starts the checking algorithm that detects the "buggy" features. The rule-based check activity the necessary data is fetched from the database, checked against a set of quality checking rules; some fix on the data is applied and exported back to the database.

The Use Case diagram of Figure 1, helps to identify the necessary software components for each system's feature. In summary, we need:

- A very flexible database management component that can effectively deal with spatial information
- Set of tools to efficiently import OSM and its updates into the database
- A powerful OGC compliant Web Service engine equipped with graphical renderer and the necessary CRS conversion means
- A rule-based quality check engine
- A workflow engine to orchestrate the frequent updating activities .

The Detailed Design

Data and Tools Diagram

Figure 2 shows the current configuration of the tools and the data flow inside the designed system.

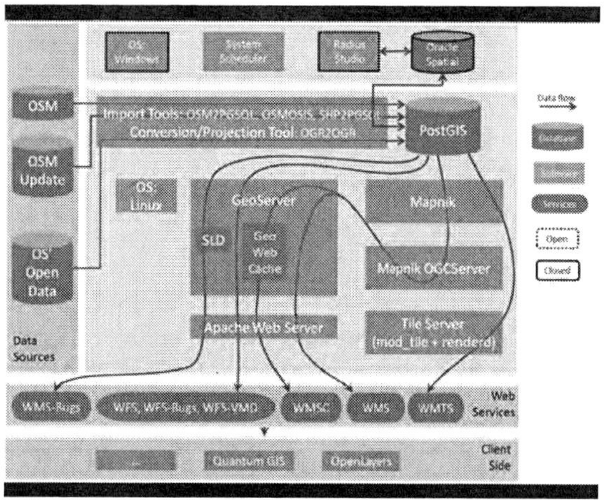

Figure 2: System Architecture; open tools, data and services.

As shown in figure 2, the current system mostly comprises open-source tools for its primary databasing and web services provision. It also shows that the data flow starts from the sources of open data and ends with the delivery in open standards and formats.

Open-source Tools and Databases

In the following subsections, the choices of open-source tools and databases are discussed. For each item, any available proprietary solution will be discussed.

PostGIS: The system databases are hosted in PostGIS 8.4.13 on Linux Ubuntu 11.0.4 platform. It has provided a very robust and stable functionality during the past year since the start of the project. The hosted databases include the whole GB processed data, daily differences, analysis data and the Ordnance Survey's VectorMap District. The first database contains just about 3 million line features, 2 million polygons and 1 million Points of Interest (POIs). During the past years, although the daily workload of the database has been relatively heavy, there was not even one case of database crash.

The main alternative to PostGIS would be Oracle Spatial. We have had an instance of Oracle Spatial 11g available. However the limitations are firstly the efficiency of importing data from OSM-XML (using either open or proprietary tools), and secondly the efficiency of running spatial queries. In both cases, PostGIS has shown a very much better efficiency when tested. Moreover, the other limitation in adopting Oracle Spatial lies in the available OSM renderers.

OSM2PGSQL: The whole OpenStreetMap data file (also called the planet file) are available in OSM-XML format through a number of OSM mirror sites around the globe including Geofabrik (Geofabrik 2012). The data in this format includes all nodes, ways (a term used for lines and polygons in OSM) and relation features together with all of the associated attributes for each feature. On the other hand, the minutely, hourly and daily updates of the planet file are available in OSC (OpenStreetMap Changes) format via the OSM replication website (PlanetOSM n.d.). The OSC format is very similar to OSM-XML but excludes unchanged features, and the changed features are separated by the update type (create, change or delete). Each OSC file includes the change from its last replication time, whether a minute, an hour or a day ago.

OSM2PGSQL (OSM-Wiki 2012) is an open-source tool that can import OSM-XML files into PostGIS database. It is also used to do the database update using OSC files. Its usability via the command line has made the simple documentation for the utility sufficient. In addition, the command-line usability has made this tool suitable to fit into workflows execution, compared with GUI-based alternatives.

There are some other tools that can import OSM-XML files into databases. OSMOSIS is an open-source tool that can do the import from OSM-XML to PostGIS but the database it produces is not directly usable by Mapnik.

Go Loader (Snowflake 2013) has been a non-open alternative. This tool has a recently added functionality that can import OSM-XML into Oracle Spatial database. However the process time is more than double of what is required by OSM2PGSQL. Moreover, it cannot work with OSC files so far. OSM2PGSQL has shown a very good performance compared with the alternative means studied for importing OSM-XML and OSC files.

Mapnik Library and OGCServer: Mapnik is the open-source tool for rendering the OSM data and is the default tool used by the OSM website. It uses the data stored in PostGIS and produces graphical output using a defined stylesheet. OGCServer is another open-source tool based on the core Mapnik library. It can make a server in order to wrap the graphics rendered by Mapnik in standard Web Services. We used the WMS output of this tool, which is by default able to make 90 separated raster layers out of source OSM data.

The main limitation we had in working with Mapnik was the documentation. For setting it up and for any customization, it has been difficult to find a complete documentation source. Individual developers' blogs, online forums and email lists have been mainly used, with the drawback of longer development and debugging time.

The alternatives have been very limited. OsmaRender is another open-source tool but it is no longer maintained, is even less documented than Mapnik and also is less used by the developers because of its known projection bug (OSM-Wiki 2012). TileMill by MapBox (MapBox n.d.) is a more user friendly cartographic system in which users can define their own stylesheet in a graphical user interface. TileMill in fact uses Mapnik as its renderer core. We therefore decided to use Mapnik in this project to reduce any unnecessary wrapping or overhead around the core rendering engine, particularly because we did not need to customized the rendering beyond the Mapnik defaults used on the OSM website.

Mod_Tile module and Renderd: These two open-source tools (OSM-Wiki 2012) are the front-end and back-end of the system's tile server (WMTS). They have nothing to do directly with the Mapnik OGC-Server, since they serve map tiles directly through a different port.

Mod_tile is an Apache Web Server module that responds to tile requests and renderd is another component that handles the file system behind the tile cache. We used this combination to serve the tiles. Not only this is used in the "slippy map" front-end of the project website, but also this can be used for any client application that can use WMTS as the background map. Tirex (OSM-Wiki 2012) is another open-source alternative to renderd. Renderd is currently used by the OSM website which made it our first choice. The system will maintain the tile cache, allowing tiles to expire and be regenerated in response to data updates.

GeoServer and GeoWebCache: The Web Service functionalities are based on GeoServer. The embedded GeoWebCache component has also been used to manage pre-rendered, cached raster outputs.

GeoServer has the ability to wrap external WMSs which is particularly useful in our system. Although the Mapnik OGCServer makes a WMS server, wrapping it with GeoServer gave better service management. Particularly the WMS made by Mapnik OGC-Server includes 91 layers (90 thematic layers plus a merged one) but by wrapping in GeoServer we could grouping layers and use custom projections. As a result, the 91 layers have been organized in 8 thematic groups: each has a number of sub-layers and are served in a customized list of projections. For example, "transportation" is a WMS layer group consisting of 40 sub-layers (different road types, bridges, tunnels, ferry routes, etc.). A user can then load any combination of the sub-layers or load the whole transportation layer in one.

Moreover, GeoServer can directly access the PostGIS database. As the result, a number of other Web Services have been created including raster and vector services for OSM, the OSM detected bugs and OS VMD. In total, the services produced by GeoServer include 2 WMSs, 1 WMSC, 1 WMTS and 3 WFSs.

We have started with GeoServer version 2.1.2 and later upgraded it to version 2.2.3. The issues we had in the older version was frequent crashes and memory leaks, limited projections and some bugs in GeoWebCache. Upgrading to 2.2.3 has shown a better stability, more integration and full support of the

BNG coordinate system in GeoWebCache. There are other open-source alternative for Web Services e. g. MapServer. GeoServer has been chosen for the ease of access and customization through its Web interface.

OSMOSIS: This open-source tool is a very general processing tool for OSM data (OSM-Wiki 2012). The pipe-line design of this command-line tool can do a variety of different jobs in importing, exporting, querying and filtering OSM data. In our system, this tool is used as a part of database updating process. OSMOSIS extracts and integrates the OSC (change files) from the OpenStreetMap replication website and makes it available to be used by OSM2PGSQL. This tool has also shown a stable performance during the project lifetime.

OGR2OGR and GDAL Library: This is an open-source command-line program that uses the GDAL library to import and export between different spatial file formats and coordinate systems (GDAL n.d.). We have used this to convert the coordinate reference systems when doing the import/export jobs for the PostGIS databases.

SHP2PGSQL: This is another open-source command-line program that comes with PostGIS installation. It converts ESRI's shape files into PostGIS database. This tool is used to import the Ordnance Survey's VMD into the database. VMD is openly published on CD-ROM as a series of shape files separated by the National Grids. SHP2PGSQL together with OGR2OGR is used to make a single database of the VMD in PostGIS that is used to serve the data through WFS.

Client Side Applications; OpenLayers and Quantum GIS: Although the standard outputs can be used by any OGC compliant client application, the focus of our system testing has been on two open-source tools: OpenLayers as the web interface tool and Quantum GIS as the desktop application. Both applications are able to work with WMS, WMTS, WMSC and WFS outputs. However, we mainly recommend OpenLayers for the light-weight services (WMTS and WMSC) and Quantum GIS for the heavier jobs (WMS and WFS).

We found a couple of issues in adopting these two applications with the other system's components: firstly in order to work with WMSC, OpenLayers uses the default value of 72 dpi for resolution (OpenLayers n.d.), while GeoServer uses OGC's 90dpi standard (GeoServer 2013). This shall be resolved by enforcing 90dpi in OpenLayers, otherwise the maps are not shown in the right position.

Secondly, requesting large vector maps can easily be very resource consuming. It is then useful to have an option of geometrically limiting the bounding box in a WFS request. In the WFS plug-in of Quantum GIS version 1.7 there is a checkbox for "Only request features overlapping the current view extend". However we found this option disappeared in the integrated WFS in version 1.8 which makes it impossible to work with large WFS datasets, unless the bounding box values are entered manually in the "WFS Filters" section.

Non-open-source Tools and Databases

As mentioned before, open-source tools have been a preferred but not a necessity when there is no better alternative. The sub-system of quality check and fix has a number of proprietary pieces of software. This has been because of the limitations in existing open-source software.

Radius Studio (1Spatial 2012) from 1Spatial is a powerful spatial information handling tool that has started its development since mid 90's. The established geo-processing power of this tool, especially on high computing power platforms, has been largely accepted within the community of professional spatial data users and it is used by a number of national mapping agencies, for example. Radius Studio internally uses a powerful object-oriented database that gives it the high performance in heavy geo-processing workloads. Moreover, all the system functionalities are web based and can be invoked by Web Services. In our view, there is a corresponding gap in the open source ecosystem, something that might be of interest to the open-source community.

Consequently, using a proprietary tool usually forces the utilization of some other proprietary tools. In this case, the Radius Studio's preferred OS is Windows and its preferred database is Oracle Spatial. As the result, we have been forced to have a separated server and database for the quality checking subsystem. Some parts of the two databases need to be synchronized and the proper conversion tools need to be used. The open-source OGR2OGR tool (mentioned earlier) can push the appropriate data between the two databases.

The Update Workflow

An interesting lesson learnt is in the central management of the two server sides (open-source and proprietary). We have Windows Server with proprietary tools and database on one side, and a set of open-source OS, tools and databases on the other side. On which system should the system orchestrator be located?

This component shall be able to invoke different processes on the two sides. In our experience, external invocation of the open-source components is much easier than for the proprietary ones, at least for the tools we have used on the two sides. Thus our solution has been to locate the system scheduler on the Windows server. PostGIS, the importing and conversion tools installed on the Linux Server as well as Radius Studio sessions are all called from Windows Server. The System Scheduler is in fact a series of batch scripts that are controlled by Windows' Scheduled Task application. The details of the updating workflow are illustrated in figure 3, below.

As shown in figure 3, the data update workflow, which is currently run every day, invokes a series of database tasks on both servers. In each cycle, firstly the current status of database changes from the last run is backed up and then reverted back to what is was before the last round of quality fixes. Secondly the OSM changes (in OSC format) is downloaded from the OSM replication site and applied to the PostGIS database (using OSMOSIS and OSM2PGSQL tools).

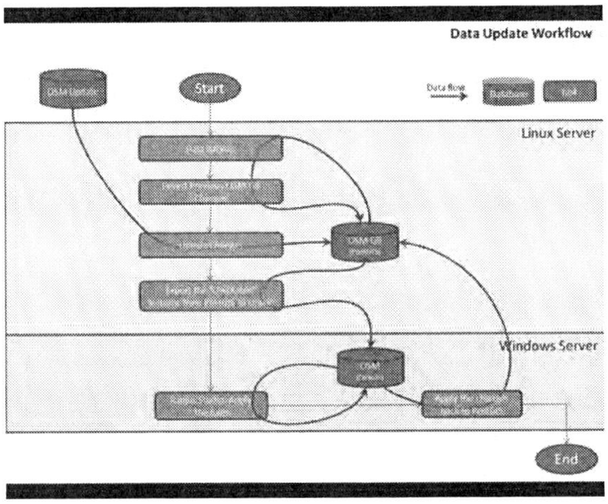

Figure 3: The data update workflow.

After the database is synchronized with the planet file, the database is ready to be processed according to the data quality rules. However, the database for GB is too big to be processed in less than a day and most of the data, which have not been updated, has already been processed in the previous run. Thus an algorithm (as PostGIS SQL commands) is run to select a subset of the OSM database which is needs to be processed by Radius Studio. This algorithm has been optimized over the life of the OSM–GB project. Briefly, this subset includes all the added and changed features, all the features that have been "buggy" in the last run, and finally all of the features in a specified proximity of those two feature groups (currently a 20 meters buffer).

After the subset is selected, OGR2OGR is used to export the subset to the Oracle Spatial database. Radius Studio then runs a processing session and tags and/or fixes the buggy features. The results of the bug detection are later applied back to the PostGIS database, again using the OGR2OGR tool. The workflow timing depends on the hardware capacities, current size of OSM in the geographical extend, the amount of daily updates and the complexity of the applied rules and actions. Currently the workflow takes about 2 hours every day and there are about 100 thousands bugs reported and/or fixed in each cycle.

Open Data

The system is an example of both using and delivering data under open licenses. The data sources and the output data of the system are generally fit under the "open" term: the OpenStreetMap data and its regular updates are licensed under Open Data Commons Open Database License (ODbL), the Ordnance Survey's Open Data (OS 2013), of which VectorMap District is a part, is licensed under the UK's Open Government License for Public Sector Information. Finally the project shares its output under Creative Common's (CC BY-SA 2.0) license.

Open Standards

The system is also a use-case of adopting the open standards from the Open Geospatial Consortium (OGC 2012). All the Web Services standards are provided in all three of the EPSG:4326, EPSG: 900913 and EPSG:27700 coordinate systems, in order to satisfy the user's requirements described in section 2. The full list of open standards used for the Web Services are:

WMS *(Web Map Service):* 90 transparent raster layers are categorized in 8 thematic groups and are served individually or in groups. The thematic groups are Land (5 sub-layers), Water (12 sub-layers), Buildings (7 sub-layers), Power (3 sub-layers), Boundaries (4 sub-layers), Transportation (38 sub-layers), Places (11 sub-layers) and Amenities (6 sub-layers). Besides the main WMS service, another WMS (called WMS-for-bugs) serves the detected errors during the quality checking. The bugs are rendered and labeled according to the type of the detected bug. An issue in WMS was the compatibility between 1.1.1. vs 1.3.0 versions. The two versions specify the x and y coordinates in opposite order, so in our experience it is better the clients specify version 1.1.1 in the WMS request in order to get the right map positions.

***WMSC** (Web Map Service – Cached):* This is a cached and single-layer version of the above WMS that optimizes the map delivery performance by keeping the pre-rendering tiles in a web cache.

***WMTS** (Web Map Tile Service):* The single-layer 256 x 256 pixel tiles are served in OGC's WMTS standard request/response. The URL of getting a single tile contains z for zoom level and x, y for the tile's relative positions in the specified zoom level (e. g. http://www.osmgb.org.uk/osm_tiles/8/126/87.png).

***WFS** (Web Feature Service):* This service makes the comprehensive vector information of OpenStreetMap in three layers (line, polygons, POIs). All the OSM key-value pairs for each feature are also included in the WFS response, which makes free filtering and querying possible. In addition to serving OSM, two other vector services exist: a WFS-for-Bugs which specifically serves the buggy features attributed with a description of each bugs, and a WFS-for-VMD which serves the imported OS VectorMap District data (for comparison purposes).

***SLD** (Styled Layer Descriptor):* This OGC standard is used to compose the "Style" files in XML, used in rendering the buggy features in WMS-for-Bugs.

***GML** (Geography Markup Language) and **KML** (Keyhole Markup Language):* These two OGC standards are among the variety of file formats used to encode the Web Services output, when applicable.

A Case Study

In this section a sample application of the usage of the developed service in an authoritative context is presented.

Surrey Heath Borough Council (`http://www.surreyheath.gov.uk`) uses a combination of Web mappings to manage its service delivery in the borough. The internal web mapping portal superimposes various map layers in British National Grid, including the maps sourced from Ordnance Survey. The WMS explained in this paper is readily used in this portal, particularly because it is available in BNG. Apart from being free, the benefit of using this layer is the "immediate" and "updateable" nature of OpenStreetMap. When new constructions are developed in the area they can be immediately reflected in OSM and updated in less than a day. Moreover, the portal users can actively contribute to this update process, because they have access to other sources of information about the county's development projects.

Figure 4 shows overlaying the "waste collection points" on top of a national Ordnance Survey's MasterMap layer. As can be seen, many waste collection points are not associated with any house because the base map layer is not current with local house construction. Figure 5 on the other hand, shows the OSM–GB WMS layer used as the base map where the new houses are already mapped (centre and upper left). Thus the waste management team is using this map base to digitize new waste collection points.

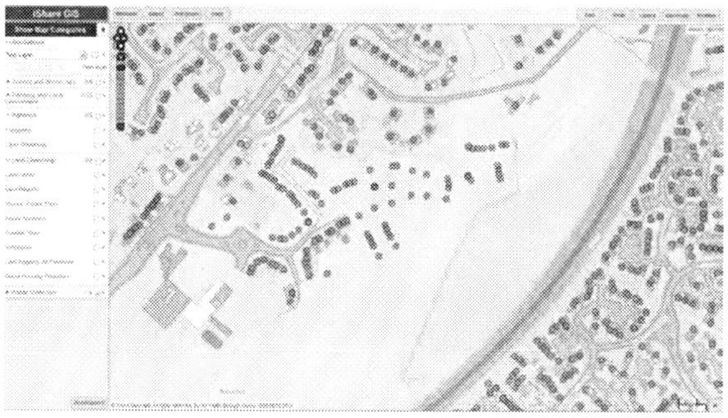

Figure 4: Waste collection points (green and blue dots) on top of MasterMap.

Figure 5: Waste collection points (green and blue dots) on top of OpenStreetMap's WMS.

However there is an area in the OSM base map (lower right) that the houses are not mapped (though they are not new houses since they are already in MasterMap – Figure 4). In such a scenario, this is a very good motivation for the portal users to voluntarily contribute to OSM and map those houses.

Benefiting from the OSM short updating cycle is the core of what this case study can demonstrate. Particularly integrating the bug layers to the GIS application, can flag up the potential errors which can be fixed manually on the OSM if necessary, so the background maps can be easily updated. Moreover, it also demonstrates addressing the other requirements described in section 2: In terms of the data adaptation, OSM data is projected in the British National Grid so it can be easily integrated to the rest of the map sources in the system, either in raster or in vector. As far as OGC compliance concerns, the WMS/WFS services have made the OSM maps one of the many layers in the employed GIS application. Finally, the multiplicity of layers in WMS allows the user to focus on the required one (e. g. buildings or transportation here). More details of this case study can be found in (Rutter 2012).

Conclusion and the Future Works

In this paper we have presented an integrated open-source, open-data and open-standards solution for OpenStreetMap utilization in a professional context. A software engineering approach has been taken to gather the professionals' requirements from OpenStreetMap and to develop the solution having as much open-source components as possible. As evidenced by case studies, it has been shown that the current open-source tools that are already developed around OpenStreetMap can put together and configured to satisfy the professional's requirements. Apart from a few practical difficulties, the open-source tools has well been integrated, however there have been certain components like quality check and improvement that a non-open-source tool has been the only available choice.

The system has many rooms to expand as the future works. As with many other software life cycles, the requirements are dynamic and the system shall address them. Optimizing the performance, expanding the quality checking rules/actions and working closer with national map agencies are examples of those future works. Another future work is for the open-source geospatial community as suggested in the paper, to consider work on a rule-based geoprocessing tool for large-scale quality check/fix purposes. We also plan further work to promote the use of the bug reporting services in the OSM community for data quality checking.

Acknowledgement This research is based at the Nottingham Geospatial Institute at the University of Nottingham, and is funded and supported by 1Spatial and KnowWhere. We also wish to acknowledge the collaborations from Ordnance Survey GB, Snowflake Software and Pitney Bowes. The inputs and helps from James Rutter in Surrey Heath Borough Council are also highly appreciated.

References

1Spatial. (2012). 1Spatial Website. Retrieved December 2012, from http://www.1spatial.com.

GDAL. (n.d.). ogr2ogr. Retrieved January 2013, from http://www.gdal.org/ogr2ogr.html.

Geofabrik. (2012). Geofabrik. Retrieved January 2013, from http://www.geofabrik.de/.

GeoServer. (2013). GeoServer 2.4.x User Manual - WMS Vendor Parameters. Retrieved January 2013, from http://docs.geoserver.org/latest/en/user/services/wms/vendor.html.

Girres, J. F. and G. Touya (2010). Quality assessment of the French OpenStreetMap dataset. Transactions in GIS 14(4): 435-459.

GIScience, U. o. H. (2010). WebMapService of World with OpenStreetMap-Data. Retrieved January 2013, from http://www.osm-wms.de/.

Haklay, M. (2010). How good is volunteered geographical information? A comparative study of OpenStreetMap and Ordnance Survey datasets. Environment and planning. B, Planning & design 37(4): 682.

JOSM. (2012). Java OSM editor Retrieved December 2012, from http://josm.openstreetmap.de

MapBox. (n.d.). TileMill. Retrieved December 2012, from http://mapbox.com/tilemill/.

Mooney, P. and P. Corcoran (2011). Integrating volunteered geographic information into pervasive health computing applications. Pervasive Computing Technologies for Healthcare (PervasiveHealth), 2011 5th International Conference on, IEEE.

OGC. (2012). Open Geospatial Consortium Web page Retrieved December 2012, from http://www.opengeospatial.org/

OpenLayers. (n.d.). OpenLayers JavaScript Mapping Library Documentation. Retrieved January 2013, from http://dev.openlayers.org/releases/OpenLayers-2.12/doc/devdocs/files/OpenLayers-js.html.

OS. (2013). Ordnance Survey OpenData. Retrieved January 2013, from http://www.ordnancesurvey.co.uk/oswebsite/products/os-opendata.html.

OSM–Wiki. (2012). OpenStreetMap Project Wiki. Retrieved December 2012, from http://wiki.openstreetmap.org/wiki/

OSMGB. (2012). OSM–GB Project Homepage - Measuring and Improving the Quality of OpenStreetMap for Great Britain. Retrieved December 2012, from http://www.osmgb.org.uk.

Pavlenko, A. (2011). Mapnik. Retrieved December 2012, from http://mapnik.org/.

PlanetOSM. (n.d.). Planet OSM. Retrieved January 2013, from http://planet.osm.org.

QuantumGIS. (n.d.). Quantum GIS Desktop Project. Retrieved December 2012, from http://www.qgis.org/.

Rutter, J. (2012). Is OpenStreetMapGB Useful? Damn Right it is! Retrieved January 2013, from http://jamesrutter.posterous.com/is-openstreetmapgb-useful-damn-right-it-is.

Snowflake. (2013). Go Loader. Retrieved January 2013, from http://www.snowflakesoftware.com/products/goloader/.

Zielstra, D. and A. Zipf (2010). A comparative study of proprietary geodata and volunteered geographic information for germany. 13th AGILE International Conference on Geographic Information Science.

Using Free and Open Source GIS to Automatically Create Standards-Based Spatial Metadata in Academia

First Investigations

by Claire Ellul[1], Nart Tamash[1], Feng Xian[1], John Stuiver[2] and Patrick Rickles[1]

1: University College London, United Kingdom; 2: University of Wageningen, The Netherlands. c.ellul@ucl.ac.uk

Abstract

The importance of understanding the quality of data used in any GIS operation has increased significantly as a result of the advent of Free and Open Source (FOSS) tools and Open Data, which in turn have encouraged non-specialists to make use of GIS. Metadata (data about data) traditionally provides a description of this quality information and permits data curation, but it is frequently deemed as complex to create and maintain. Additionally, it is generally stored separately from the data, leading to issues where updates to the data are not reflected in the metadata and to users not being aware that metadata exists. This paper describes an approach to address these issues in an academic context – tightly coupling data and metadata and automating elements of standards-based metadata creation and automating keyword generation and language detection. We describe research into the potential of the FOSS packages Quantum GIS and PostGIS to support this form of metadata generation and maintenance.

Keywords: Keyword1, keyword2.

1 Introduction

Advances in positioning, web mapping, mobile communications, Web 2.0 and Volunteered Geographic Information (VGI) (Goodchild 2007), along with the emergence of the Open Data movement, have led to increasing availability of spatial data (Budhathoki et al. 2008), with much of this data available free of charge (Coleman et al. 2009). The availability of free Geographical Information Systems (GIS) software (e.g. Google Earth, ArcGIS Explorer, Quantum GIS) encourages nonspecialist users to make use of GIS tools and data.

In academia, this increase in available data and software, along with the requirement to curate the data, is coupled with a reduction in GIS expertise of the end user of such tools. Given this, having information to allow end-users to understand, manage and integrate the heterogeneous data they are using, and identify any limitations, becomes more important (Deng & Di 2009, Haklay & Weber 2008).

Traditionally, among GIS professionals, metadata ('data describing the data') has been created to curate data (Sboui et al. 2009). It details how data was derived, why it was captured, at what scale and how it has been processed, covering issues related to topological correctness, semantic, temporal and positional accuracy (Goodchild 2002, Longley et al. 2011, Van Oort 2005, Burrough 1994). It provides a formal description of the data quality (Kim 1999), allows for data reuse (Craglia et al. 2008) and avoids data duplication. Good metadata increases trust (Craglia et al. 2008) and helps increase the credibility of a dataset (Coleman et al. 2009). In general, therefore, "the purpose of metadata is to facilitate the interpretation of data" (Sboui et al., 2009).

However, metadata is complex to create (Poore & Wolf 2010, Manso-Callejo et al. 2009, Batcheller 2008, Craglia et al. 2008) and is usually created by a dedicated team of professionals (Mathes 2004 in (Kalantari et al. 2010)). "Many view its generation as monotonous and time-consuming" (Batcheller 2008, p. 388). Standards are producer-centric (Goodchild 2007, Devillers et al. 2005) and quality may be variable (Rajabifard et al. 2009). Metadata production is often left to the end of a project, which results in metadata that is barely useful and often contains errors (West & Hess 2002). The current approach to data curation -where metadata is decoupled from the data it describes -further complicates this situation. Decoupled metadata may not be updated when data changes, and its existence is easily ignored by users.

This paper presents preliminary work on an approach to overcome these issues in the context of academic research and data curation. Using Free and Open Source (FOSS) GIS products -Quantum GIS 1.8.0 and PostGreSQL 9.2 with Post-GIS 2.0 (to maximize potential uptake amongst academics without incurring licensing costs), we describe how metadata creation to the metadata standard used by INSPIRE (INSPIRE 2011b) can be, in large part, automated – in particular keyword generation and language detection. Importantly, this is done in a manner that tightly couples metadata and data.

The remainder of this paper is structured as follows: firstly, we briefly outline the importance of spatial data infrastructures and metadata in an academic context, considering the relevance of INSPIRE and the ISO 19115 standard, approaches offered by current vendors and previous attempts at automation. This is followed by an investigation into the automation potential of individual elements of ISO 19115 and a description of the system architecture used and implementation approaches taken. Results are presented, particularly for language and keyword automation and the paper concludes with a discussion and an overview of further work to be carried out.

2 Background

2.1 The INSPIRE Project

The INSPIRE (INfrastructure for SPatial InfoRmation in Europe) directive, issued by the European Union in 2007 (INSPIRE 2011a), sets up a framework for the creation of an European Spatial Data Infrastructure (ESDI), which will enable the sharing and comparison of environmental information among public sector organizations and facilitate public access to spatial information across Europe (INSPIRE 2011a). Data themes covered by INSPIRE are wide-ranging and include coordinate reference systems, addressing, administrative units, land cover, elevation, environmental monitoring facilities and natural risk zones.

As with any Spatial Data Infrastructure, metadata forms a core component of INSPIRE. For INSPIRE this is based on the ISO 19115 standard ISO (2003) (referred to as "INSPIRE metadata" in this document). Core elements of INSPIRE metadata cover resource identification, keywords, geographical location, temporal references, quality and validity of the data and information about the metadata itself (INSPIRE 2011b) along with issues relating to sourcing the data and licensing its use, as well as logical consistency (the degree to which the contents of the dataset follow the specification rules), completeness (are there gaps or missing data), positional accuracy, and lineage (how the dataset was acquired or compiled) (Goodchild 2007). Indeed, a total of 38 separate items of information can be identified (INSPIRE 2011b).

2.2 Academic Context

The increasing availability of software and data is particularly relevant for many of the multi-disciplinary academic projects in which the authors of this paper are involved and which provide a motivation for the research described here. The power of GIS as a tool for the integration of data from diverse sources and disciplines means that it is frequently used in such projects. These projects, in turn, generate additional data. Curation of this data is an increasingly important area for academics, and is now mandated by funding bodies including the European Union FP7 FP7 (2011), the UK Environmental and Physical Sciences Research Council EPSRC (2011) and Economic and Social Research Council ESRC (2010). However, many academics do not have the skills required for such curation -and indeed may come from non-GIS disciplines as diverse as tourism studies, coastal geomorphology, anthropology, architecture and urban studies (Ellul et al. 2012).

Although it is as yet unclear whether the INSPIRE directive is specifically applicable to academia, and if so to what extent (Reid 2011), the general requirement to curate research data will most likely result in a requirement for the creation of standards-based metadata for academic datasets, to ensure interoperability and facilitate data exchange.

2.3 Metadata and GIS Software

Given that metadata has long formed an important element in the process of managing spatial data, and it is perhaps not surprising that many GIS packages provide functionality to create and maintain metadata as part of their functionality. The options offered by key packages are summarized in Table 1, along with metadata management tools provided by the INSPIRE project.

Limitations of Current Approaches

As indicated in Section 1, there are a number of issues with the current approaches. Firstly, the complexity of standards-based metadata means that users are not inclined to create or maintain it and therefore curate their data. Given the intricacy of metadata standards, even for specialists the complexity of creating and maintaining metadata is considered significant (Poore & Wolf 2010, Manso-Callejo et al. 2009,

Table 1: Summarizing Existing Approaches to Metadata Handling.

Package	Summary
ESRI ArcGIS 10.1	Metadata in ArcGIS is created via the separate catalog tool which lists available layers and datasets. Right clicking on a dataset opens up a properties window which allows the user to enter information including a title, tags, a summary, a description of the dataset, credits and access and use limitations. Metadata can also be imported from, or exported to, standards-based XML. It is embedded within the shapefile format, which means that it persists between projects. As the metadata is embedded, it is not possible to search through multiple metadata records unless these are exported.
Geomedia Professional 6.1	Geomedia Professional offers a 'catalog' tool for metadata creation and management, along with the capability to import existing metadata and export metadata for use in other systems. Catalog records are ISO 19115 compliant. Catalogs are stored as Microsoft Access databases (.mdb), decoupled from the datasets. The catalog creation process will automatically populate information including the bounding box in the Catalog Editor tool. No functionality is provided to search all created metadata for specific keywords or themes.
Quantum GIS 1.8.0	In QGIS two alternative metadata options are offered. Users can create simple metadata directly with the properties of each dataset. The metadata is stored in the system project file (and therefore not available to other projects or users making use of the same datasets). Alternatively, a plug-in is also available - a metadata editor called Metatools, which can read and write metadata in ISO19115 format. The main purpose of this tool is to create standards-compliant metadata within QGIS, primarily for export to HTML format. This metadata editor is separate from the main workflow of a QGIS user (Lab n.d.). No functionality is provided to search all metadata.
INSPIRE Portal	The INSPIRE Geoportal (The INSPIRE GeoPortal n.d.) provides a central viewer for any available metadata created as part of the INSPIRE project. Unlike the stand-alone GIS packages, it provides a search tool for all created metadata, where searches can be spatial or text-based. Additionally, a metadata editor tool is provided to allow users to create INSPIRE compliant metadata. This can be validated using the tools provided. The metadata is held separately from the datasets referenced.

Batcheller 2008, Craglia et al. 2008). Secondly, metadata is, in most cases, de-coupled from the related dataset and in all cases does not form an integral part of the user's workflow when opening or editing data inside the GIS. This has two consequences -it is possible for users to use a dataset without being aware of any limitations or constraints -issues that are particularly relevant for novice users. It is also possible to edit and change the data without updating the corresponding metadata or to maintain metadata in one GIS but not make it available automatically to users of another GIS. This is particularly important to support interoperability.

To address these issues, metadata should be more closely coupled with the data itself and its creation should be as automated as possible. Where this is not possible metadata creation, maintenance and use should be integrated into the user's workflow. The remainder of this paper describes a first investigation into the potential of FOSS GIS to achieve these above aims.

3 Automating Metadata Creation

3.1 Previous Work

As has been seen (Section 2.3), a number of metadata elements are already created automatically by the various GIS packages. These include the identification of the bounding box coordinates of a dataset and the relevant projection or reference system. Beyond these basics, (Kalantari et al. 2010) have introduced a framework for the spatial metadata enrichment. Their work examines the potential of using concepts relating to tagging and folksonomies (collaborative tagging). Based on searches against a metadata repository, they assign the user's search words as keywords to any datasets that the user downloads as a result of a search process, and also propose direct user-tagging approaches to enriching metadata. (Olfat et al. 2012) introduce process-based metadata entry, which creates metadata in parallel with the dataset life-cycle, rather than after the generation of dataset or at the end of the project. They propose the coupling of metadata and data in one database. Their architecture is web based, making makes use of GML as a data transfer standard to sup-

Table 2: Potential Automation of some INSPIRE Metadata Elements.

Metadata Element	Automation Potential	Mandatory	Implemented
Resource Title	Inserted manually. If not inserted by the user, default value is the dataset name.	Yes	Yes
Resource Abstract	Inserted manually	Yes	Yes
Identifier code and namespace (2 elements)	Take the Object Identifier of the spatial table in the database. This will form part of a unique URI for the dataset, which will also incorporate project and end user domain detail. In an academic context, a default value for a project or university could be used.	Yes	Yes
Resource type	Default to 'dataset'.	Yes	Yes
Resource language	It may be possible to implement this using language detection algorithms provided the dataset contains sufficient text. See Section 4.3.1 for details	No	Yes
Keyword(s)	This could be implemented by concatenating all text fields of the dataset and picking the top 10 repeating words while eliminating common words. See Section 4.2.2	Yes	Yes
Bounding Box (4 elements)	Can be automatically identified from the spatial coordinates in the dataset	Yes	Yes
Date of publication	Can default to the date that data was uploaded to the system, with updates when the data is edited. Manual verification required by the end user.	Yes	Yes
Date of last revision	Default to the date the data was uploaded to the system. Update automatically any time data edited	No	Yes
Date of creation	Default to the date the data was uploaded. Manual verification required by the end user	No	Yes
Limitations on public access and conditions of use (2 elements)	Given the academic context, a default value can be assigned, perhaps taking the most open value or perhaps on a per project basis.	Yes	No
Responsible party name, email and role (3 elements)	Based on user groups (identified from the user's login details and a corresponding lookup table).	Yes	Yes
Metadata contact name, email and date (3 elements)	This can be derived from the database login of the person uploading the dataset or creating the new dataset.	Yes	Yes
Metadata language	This can be detected by applying a language detection algorithm to the metadata (see 4.3.1 below)	Yes	Yes

port interoperability. Loose coupling is achieved by means of a layer of middleware (Olfat et al. 2012).

3.2 Automating INSPIRE Metadata Creation

A review of the metadata standard used by INSPIRE reveals that the population of a significant number of elements can be automated, in particular when the standard is applied in an academic context (which means that pre-defined, project specific, values can be used for some metadata). Indeed, it may be possible to automate the population of the majority of the mandatory elements of the standard.

Table 2 outlines suggested approaches mandatory metadata elements (a full list, showing all metadata elements, can be found at `http://www.mapmalta.com/FOSS4G2013_FullTables.pdf`). The Table also describes which elements have been implemented in the prototype system described below (Section 4). At this point, it is important to note that, to work towards interoperability by allowing both data and metadata to be read by multiple GIS, it is assumed that all spatial data will be stored in a spatial database along with the corresponding metadata.

4 Implementing Metadata Creation in FOSS GIS

4.1 System Architecture

The approach described here builds on the concept of closely coupling metadata and data presented in (Ellul et al. 2012, Olfat et al. 2012). However, unlike Olfat et al. (2012), the coupling in this case takes place via triggers embedded in the database itself rather than relying on middleware. Triggers are automatic functions that run whenever data is inserted, updated or deleted in a database table, providing a very tightly coupled relationship between data and metadata, and their presence means that metadata is automatically updated when the dataset is edited in any way. Figure 1 shows the overall system architecture.

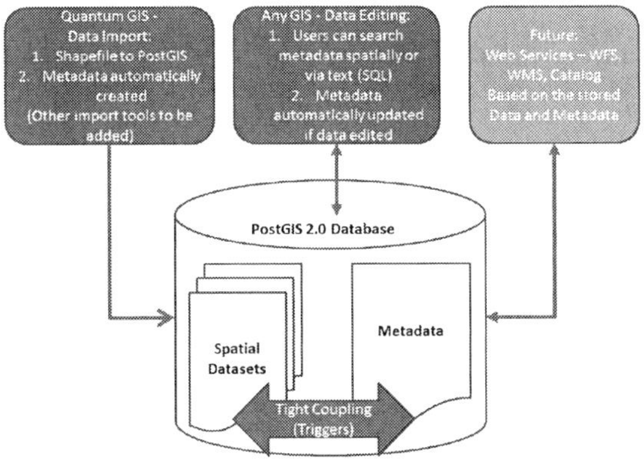

Figure 1: System Architecture.

The system is built using Quantum GIS 1.8.0 and PostgreSQL 9.2 with Post-GIS 2.0. The database was selected due to its interoperability with multiple GIS packages -including ArcGIS and Geomedia Professional. Quantum GIS offers the option to develop plug-ins using Python, and also offers excellent interoperability with the selected database.

4.2 Configuring the PostGIS Database and Triggers

To support metadata automation, two tables have been created in the PostGIS database -firstly, the metadata table itself and secondly, a table created to store required user information (based on the user's database login details) to automatically populate metadata when data is uploaded or modified. This 'database groups' table lists the group name (which corresponds to the PostGIS login group), the contact e-mail for that group and the user's organization name.

4.2.1 Database Triggers

Triggers are created in PostgreSQL's inbuilt programming language -PL/pgSQL (*The PL/pgSQL Programming Language* n.d.). They add the following metadata details (see Section 4.3 below):

- The dataset extents, taken from the spatial extents of the data, using an ST_Envelope query
- The resource title, taken from the table name
- The Identifier Code, taken from the Object ID in the database
- The last revision date -defaults to the current date and time
- The metadata contact details -extracted from the PostGIS user's login details
- A bounding box geometry for the metadata, taken from the dataset extents
- The metadata date -taken from the current date and time
- The responsible party -taken from the PostGIS login
- Keywords -extracted from the dataset text (see Section 4.2.2)

A final procedure is then run to create a trigger on the newly uploaded dataset. This will automatically update the metadata every time the dataset is changed.

4.2.2 Identifying Keywords

The PL/pgSQL code used to identify keywords first identifies any text fields in the data table, and then splits the text into single words using a space as a delimiter. A UNION operation is used to generate a long, one-column, list of the words, and a *group by* query then used to identify the 10 most common words and known common words (*yes, no, or, and*) are eliminated from the list.

4.3 Writing the QGIS Plug-Ins

When using the Data Loading and Metadata Creation QGIS plug-in the user first selects a data file (shapefile) to load into the database, and an automatic connection to the database is established. The user can rename the file, and must type in the required metadata into the form -i.e. the title, abstract and, if available, lineage information. Once the user presses the 'OK' button, a Python process is run to load the data into the database and then insert the appropriate metadata.

The Python code first creates the spatial table in the database, naming the table with the filename of the uploaded file. The field names are identified from the shapefile and a geometry column of the appropriate type (again, identified from the shapefile) is added to the table. The SRID (spatial reference ID) value is taken from the form. Following this, the Python code iterates over the shapefile and inserts the data into the new table, row by row. Finally, the metadata language and dataset languages are identified (see Section 4.3.1 below), and required metadata inserted into the database. This metadata creation process -specifically the INSERT activity into the metadata table -triggers the PL/pgSQL procedures described above (Section 4.2.1) -i.e. the automated metadata creation.

4.3.1 Identifying the Metadata and Dataset Language

Both metadata and dataset language identification processes make use of a Python library known as "langid" (Lui & Baldwin 2012, n.d.). This language detection tool, based on machine learning algorithms (Lui & Baldwin 2011), is configured for 97 different languages, and works by a process of pre-training, in which common tokens for each language are identified through the examination of a set of documents in the given language. Importantly, it has been developed to allow cross domain applicability -for example, if the process has been trained to recognize Italian using a series of documents relating to air quality, it should be able to adapt this training to other Italian documents in a different domain. To detect the metadata language, three pieces of text are concatenated -the title, abstract and, if available, the lineage. For the dataset language, the first 10,000 characters of text are concatenated as the program is iterating through the dataset.

5 Testing Metadata Automation

In order to test the various key elements of the metadata automation process -and in particular keyword generation and dataset language detection, a range of Open Street Map datasets (OSM) (Haklay & Weber 2008) from ten different European countries were selected and uploaded into PostGIS using the plug-in described above (Section 4.3). OSM data was selected as it provides identically structured, multi-lingual (at least in part) data from around the world. This permits extended testing for keyword creation and language detection. An element of data cleaning has been carried out by Cloudmade. For each country the roads datasets ('highways'), location datasets (which detail key locations in each country), the points of interest datasets and the administrative boundary datasets were downloaded from `http:\www.cloudmade.com`. Table 3 shows an extract of the results obtained, with the number of keyword occurences given in the brackets in each case.

Table 3: Results for Open Street Map Datasets.

Dataset	ResourceLanguage	Keywords
austria_administrative	de	8(10889), 6(2158), 9(893), 10(875), 2(690), 4(527), /(371), Border(279), 7(264), StraÃ?e(203)
austria_highway	ht	track(288357), residential(265675), service(179314), path(86165), unclassified(78983), footway(78193), tertiary(32408), secondary(29048), StraÃ?e(28442), primary(24575)
austria_location	de	MÃijnchen(21504), hamlet(19230), Wien(13821), village(13086), 1(5866), Germering(5317), 82110(5312), 2(5292), Bad(4812), locality(4628)
austria_poi	en	Public(122871), Services(119379), Government(119379), Power(60686), Tower(59271), Automotive(52389), Tourism(45779), Bus(24174), Eating and Drinking(22032), Parking(20230)
greece_administrative	el	8(842), 6(317), 4(293), -(149), 2(95), Border(70), Î(60), sÄśnÄśrÄś(54), Î?ÎżÎżÎňÎt΄Îś(40), 7(38)
greece_highway	la	residential(107535), tertiary(22959), track(17127), unclassified(8819), secondary(8291), primary(6817), ?d??(6644), service(6377), footway(4517), motorway(3605)
greece_location	qu	village(6125), hamlet(3021), ???????(1344), ????????(1225), ??????(1130), ?????(1103), ?????????(884), ??????????(526), ????(521), town(298)
greece_poi	en	Public(20481), Services(20226), Government(20226), Tower(14798), Power(14757), Automotive(7226), Tourism(4412), Pedestrian(3117), Crossing(3117), Eating and Drinking(2752)

5.1 Comparing the Results

Table 3 above gives a sample of the results obtained[7]. The results show that metadata has been correctly identified as being in the English language in all cases. However, in general both the keyword extraction process and the language identification process for the resource have yielded mixed results.

Firstly, all the points of interest data yielded 'English' as the language and terms including 'Public', 'Services', 'Tourism' and so forth as keywords. Examining the OSM points of interest datasets yields a potential explanation. Much of the data is in fact placed into category names which are given in English (no matter the country in question) – categories include 'Automotive', 'Government and Public Services', 'Tourism' and so forth. Additionally, although perhaps less expected, the types of the points are also given in English -for example in the Austrian dataset, we find 'Museum:Ortsmuseum Tutzing' and 'Significant tree' and 'Peak:Oberer Burgstall'. If the language issue is perhaps put to one side (as the data is, indeed, in English) it could be said that the key-

[7]Note that data is only shown for Austria and Greece due to space restrictions. A full listing can be found at `http://www.mapmalta.com/FOSS4G2013_FullTables.pdf`

words do provide a good representation of the Points of Interest Dataset, giving a mix of the types of information that this dataset provides.

For the administrative data, in all cases except for Belgium, the correct language was identified. In the Belgian case, however, the language was identified as 'lb' -Luxembourgish. This could be due to the fact that in Belgium both the French and Flemish languages are used. Keyword identification, on the other hand, was not as successful -indeed, numbers were identified as the most common 'words' in all cases. Examining the datasets again yields an explanation -in all the datasets, much of the 'name' data is blank (or null), whereas the 'administrative_level' data -which is a number detailing where on the overall Administrative boundary hierarchy the data element falls -is fully populated.

For the highway data, a more mixed result is noted -Malta, Italy, Spain and Greece all yielded 'Latin' as the language, Portugal and Sweden correctly yielded Portuguese and Swedish. However, the Netherlands yielded 'ms' (Malay), Belgium yielded 'jv' (Javanese) and Austria yielded 'ht' (Haitian). Examining the Belgian data, it can be seen that it is a mixture of Flemish, French and English, although English terms such as 'residential' and 'path' do dominate. Keywords in this case were given predominantly in English (due to the underlying data) and included 'track', 'unclassified', 'residential', 'footway' and 'cycleway'. Some common words in each language also made it to the list - Triq (Maltese), Via (Italian), Rua (Portuguese), Calle (Spanish), Strasse (Austria), Rue (Belgium), OdoV (Greece) which all translate as street.

Finally, for the location dataset language identification, Sweden yielded 'Danish', both Spain and Portugal yielded 'Spanish', the Netherlands and Belgium yielded 'Netherlands', Malta yielded 'Latin', Italy yielded 'French' and Austria yielded 'German'. Greece yielded 'qu' (Quecha, which is a native South American language). In this case, the latter could perhaps be attributed to problems with the Greek characters in the text, which rendered as "?" in the database, along with the inclusion of Greek place names transliterated into English (such as 'Komianata' or 'Agii Deka'). Again, keywords were predominantly in English -'locality','hamlet','village' due to the English language place data embedded in the datasets. However, these were mixed with place names (Aachen, Fgura, Birmingham, Munchen, Wein, Brugge, Trento) representing the most commonly used location points in each dataset. Provided the user understands English, the keywords do to a certain extent represent the dataset well.

To summarize the above results, in three cases out of the four tested, the keywords yielded from the datasets did provide a relatively useful list of words relevant to the dataset in question. The results for the fourth case -administrative boundaries -could perhaps be improved upon by eliminating keywords having very short length, or consisting of numbers, from the potential keywords list. Equally, it is important to identify and remove all common words ('and','or' and so forth) in the relevant languages from the list of potential keywords. This was temporarily hard-coded for the English language, but would require input from speakers of other languages to add to this list, which should then be stored in a table in the database.

Language identification also yielded rather mixed results, in particular where multiple languages were included in the dataset. A number of heuristics could be suggested to improve this process, however. A simple spatial intersection with a world map would identify the country where the data is located. This could then be used as a suggested language or languages to the language identification process. This is particularly important when two languages are close in nature e.g. Flemish, French and Luxembourgish. The possibility of multiple languages within one dataset should also be considered and accounted for in the data model.

6 Discussion and Further Work

The work described here details a preliminary investigation into the potential of automating metadata – and in particular an investigation into the potential of closely coupling metadata and data, automatically generating keywords and detecting the language used in the dataset and metadata. Despite the issues with language identification, overall the work yielded promising results. Importantly, we have shown that metadata and data can be tightly coupled so that modifying data automatically updates the metadata. We have also shown that this is possible within FOSS GIS software. By embedding the coupling within the spatial database, the functionality to maintain dataset and metadata synchronized when data is edited is interoperable across multiple FOSS and non-FOSS GIS platforms. We have been able to automatically populate a total of 18 of the 20 mandatory INSPIRE fields, with a further 4 optional fields populated (from a total of 18 optional fields).

Both datasets and metadata are stored in an open spatial format, which means that they can be shared with other GIS packages -this makes the entire metadata catalog searchable within the GIS. The central database approach also permits the data and metadata to be published as Web Feature Services and Catalog Services for the Web, providing a discovery type service similar to that used in INSPIRE. While the work described in this paper has focused on metadata in an academic context many of the approaches described above are relevant elsewhere, although it is possible that using these approaches in a more general context may reduce the number of metadata elements that can be automated.

A number of technical issues remain to be addressed -in particular, the difficulty encountered when handling Greek and other non-Latin characters in the database and potential performance issues caused when significant changes are made to the underlying datasets on a row by row basis. Potentially, the user could be given the opportunity to temporarily disable metadata updates, or to set them to run in batch mode overnight.

Increasing the interoperability of the approach also presents an interesting challenge. At this point in time, functionality developed works when a pre-created dataset is imported into the database, and when the resulting dataset is edited by any GIS the automated elements of the metadata are updated - i.e. partial interoperability has been achieved. Interoperability should, however, be extended to incorporate any new spatial table created in the database or imported via other mechanisms. Semantic interoperability of the manually-populated elements of the metadata standard also present a problem and may require the development of plug-ins for the creation of the manual elements of metadata via other GIS or import of existing metadata. It would also mean that keyword and language identification would not necessarily be immediately possible, but could instead be run as a batch process once sufficient data is added into the table. For total interoperability, the language detection algorithm should be incorporated directly into the database rather than embedded in the plug-in.

Given the mixture of languages within the OSM datasets, perhaps the next step in this research should also be to identify appropriate single-language datasets (along with corresponding metadata) to conduct further tests. Language experts for each language should also be involved to ensure that the results yielded are appropriate -in particular the keywords identified. Once this is complete, further work could continue on the other elements of metadata automation. The system could also be extended to include non-INSPIRE metadata -(Ellul et al. 2012) notes a number of elements such as tags, dataset and metadata ratings that could be relevant. Potentially, given that the aim of this tool is for use in an academic setting, additional project-related information (for example which Work Package generated a dataset) could also be added. Finally, issues relating to deployment should also be considered -these tools require a level of expertise to initially set up and configure, which could be provided by data management support staff within each academic institution.

References

Batcheller, J. K. (2008), 'Automating geospatial metadata generation [2DA?]Uan integrated data management and documentation approach', *Computers & Geosciences* 34(4), 387–398. Budhathoki, N., Bruce, B. & Nedovic-Budic, Z. (2008), 'Reconceptualizing the role of the user of spatial data infrastructures', GeoJournal: *An* International *Journal on* Geography 72, 149–160.

Burrough, P. (1994), *Principles of Geographical Information Systems for Land Resources Assessment*, Oxford Science Publications, Clarendon Press, Oxford, UK.

Coleman, D., Georgiadou, Y. & Labonte, J. (2009), 'Volunteered geographic information: The nature and motivation of produsers', *International Journal of Spatial Data Infrastructures Research* 4, 332–358.

Craglia, M., Goodchild, M., Annoni, A., Camara, G., Gould, M., Kuhn, W., Masser, I., Maguire, J., S, L. & Parsons, E. (2008), 'Next generation digital earth. a position paper from the Vespucci initiative for the advancement of geographic information science', *International Journal of Spatial Data Infrastructures Research* 3, 146–167.

Deng, M. & Di, L. (2009), 'Building an online learning and research environment to enhance use of geospatial data', *International Journal of Spatial Data Infrastructures Research* 4.

Devillers, R., Bedard, Y. & Jeansoulin, R. (2005), 'Multidimensional management of geospatial data quality information for its dynamic use within GIS', *Photogrammetric Engineering and Remote Sensing* 71, 205–215.

Ellul, C., Winer, D., Mooney, J. & Foord, J. (2012), 'Bridging the Gap Between Traditional Metadata and the Requirements of an Academic SDI for Interdisciplinary Research', *Spatially Enabling Government, Industry and Citizens* .

EPSRC (2011), 'EPSRC Policy Framework on Research Data', `http://www.epsrc.ac.uk/about/standards/researchdata/Pages/default.aspx`

ESRC (2010), 'ESRC Research Data Policy', `http://www.esrc.ac.uk/_images/Research_Data_Policy_2010_tcm8-4595.pdf`

FP7 (2011), 'Research Data Management -European Commission -FP7', `http://www.admin.ox.ac.uk/rdm/managedata/funderpolicy/ec/`

Goodchild, M. (2002), 'Introduction to part i: Theoretical models for uncertain GIS', *Spatial Data Quality* pp. 1–4.

Goodchild, M. (2007), 'Citizens as sensors: the world of volunteered geography', *GeoJournal* 69, 211–21.

Haklay, M. & Weber, P. (2008), 'Openstreetmap -user generated street map', *IEEE Pervasive Computing* pp. 12–18.

INSPIRE (2011*a*), 'About INSPIRE -Infrastructure for Spatial Information in Europe', `http://inspire.jrc.ec.europa.eu/index.cfm/pageid/48`

INSPIRE (2011*b*), 'INSPIRE Metadata Implementing Rules: Technical Guidelines based on EN ISO 19115 and EN ISO 19119', `http://inspire.jrc.ec.europa.eu/documents/Metadata/INSPIRE_MD_IR_and_ISO_v1_2_20100616.pdf`

ISO (2003), 'ISO 19115:2003, Geographic Information – Metadata'.

Kalantari, M., Olfat, H. & Rajabifard, A. (2010), 'Automatic spatial metadata enrichment: reducing metadata creation burden through spatial folksonomies', *Global Spatial Data Infrastructures 12 World Conference: Realising Spatially Enabled Societies* .

Kim, T. (1999), 'Metadata for geo-spatial data sharing: a comparative analysis', *The Annals of Regional Science* 33, 171–181.

Lab, G. (n.d.), 'Working with metadata using metatools for QGIS', `http://gis-lab.info/qa/metatools-eng.html`

Longley, P., Goodchild, M., Maguire, D. & Rhind, D. (2011), *Geographical Information Systems and Science Third Edition*, Wiley Hoboken NJ.

Lui, M. & Baldwin, T. (2011), 'Cross-domain feature selection for language identification', `http://www.aclweb.org/anthology-new/I/I11/I11-1062.pdf`

Lui, M. & Baldwin, T. (2012), 'langid.py: An off-the-shelf language identification tool', `http://www.aclweb.org/anthology-new/P/P12/P12-3005.pdf`

Lui, M. & Baldwin, T. (n.d.), 'langid.py', `https://github.com/saffsd/langid.py`

Manso-Callejo, M., Wachowicz, M. & Bernabé-Poveda, A. (2009), 'Automatic metadata creation for supporting interoperability levels of spatial data infrastructures', *GSDI 11 World Conference: Spatial Data Infrastructure Convergence: Building SDI Bridges to Address Global Challenges* .

Olfat, H., Kalantari, M., Rajabifard, A., Senot, H. & I, W. (2012), 'Spatial metadata automation: A key to spatially enabling platform', *International Journal of Spatial Data Infrastructures Research* 7, 173–195.

Poore, B. & Wolf, E. (2010), 'The metadata crisis – can geographic information be made more usable?'.

Rajabifard, A., Kalantari, M. & Binns, A. (2009), 'SDI and Metadata Entry and Updating Tools', *SDI convergence: Research, Emerging Trends and Critical Assessment* .

Reid, J. (2011), 'The EU INSPIRE Directive and what it might mean for UK Academia', `http://www.data-archive.ac.uk/media/338363/INSPIRE7Oct2011jr.pdf`

Sboui, T., Salehi, M. & Bédard, Y. (2009), 'Towards a quantitative evaluation of geospatial metadata quality in the context of semantic interoperability', *6th International Symposium on Spatial Data Quality* .

The INSPIRE GeoPortal (n.d.), `http://inspire-geoportal.ec.europa.eu` Accessed: February 2013.

The PL/pgSQL Programming Language (n.d.), `http://www.postgresql.org/docs/9.0/static/plpgsql.html`

Van Oort, P. (2005), 'Spatial data quality: From description to application', *PhD, Wageningen University, The Netherlands* .

West, L. & Hess, T. (2002), 'Metadata as a knowledge management tool: supporting intelligent agent and end user access to spatial data', *Decision Support Systems* 32, 247–264.

Gestural Interaction with Spatiotemporal Linked Open Data

by Thomas Bartoschek[1], Gerald Pape[1], Christian Kray[1], Jim Jones[1] and Tomi Kauppinen[2]

1: University of Münster, Germany. 2: Aalto University, Finland bartoschek@uni-muenster.de

Abstract

Exploring complex spatiotemporal data can be very challenging for non-experts. Recently, gestural interaction has emerged as a promising option, which has been successfully applied to various domains, including simple map control. In this paper, we investigate whether gestures can be used to enable non-experts to explore and understand complex spatiotemporal phenomena. In this case study we made use of large amounts of Linked Open Data about the deforestation of the Brazilian Amazon Rainforest and related ecological, economical and social factors. The results of our study indicate that people of all ages can easily learn gestures and successfully use them to explore the visualized and aggregated spatiotemporal data about the Brazilian Amazon Rainforest.

Keywords: Gestural interaction, spatiotemporal phenomena, Linked Open Data.

1 Introduction

In recent years, gestural interaction has been successfully used to enable untrained users to control different types of applications. Some operations arguably map naturally to certain types of gestures. Examples of this are pointing at an item to select it or moving a body part to trigger a similar motion of the screen content. These properties of gesture control are typically used to facilitate access to simple data such as photographs, media or basic maps. Given this, our hypothesis is that gestural interaction is equally well suited to enable the exploration of more complex (linked) spatiotemporal data. The goal is thus to create methods for communicating results of the field of Geographic Information Science to the public, for example in an exhibition, in a science museum or science center, where large surfaces are important for better visibility.

Gesture control is a promising approach to enable interaction with large surfaces in particular since it can improve ease of learning and help overcome reachability issues [15] and due to its high level of learnability [7, 23]. While previous work has shown this for basic spatial operations (panning, zooming, re-arranging objects), it is not clear whether gestural control can work equally well when interacting with spatiotemporal data.

Interfacing research on Human-Computer Interaction with recent Linked Open Data and visualisation techniques for complex spatiotemporal data is a contribution towards Linked Open Science [14] to support transparency and openness of science via facilitating the exploration of scientific observations. Clearly, Linked Open Science needs Linked Open Data (LOD)[8] to allow for publishing of very different kinds of data on the web, and to interconnect them together and to space and time. Additionally Linked Open Science needs Open Source Software, to make its results reproducable and freely available.

In this paper we present and evaluate an application based on open source technologies enabling the exploration of Linked Spatiotemporal Data integrated into an exhibit. We present a set of gestures that visitors to a science fair can use to explore large amounts of linked data related to deforestation on large screens and we also report on results from an initial survey. In the following, we first briefly describe the background before presenting the prototype system, the linked open spatiotemporal datasets [13], their visualizations and the gestures we have implemented. We then describe the survey we conducted and summarise the results we obtained. The penultimate section discusses possible implications of our findings on the design of similar systems. The paper closes by summarising the main contributions and giving a brief outlook on future work.

2 Related Work

There is a broad range of previous work investigating the use of different types of gestures (e.g. arm-, body-, or headgestures) to facilitate interaction with various types of systems (e. g. desktop computers, mobile phones or public displays). The recent intro-

[8] http://linkeddata.org

duction of affordable off-the-shelf solutions to track gestures [16] led to the development of gestural interfaces for a variety of application scenarios. Gestures have thus been used successfully as a means to enable laypeople to control different types of applications (e. g. simple map exploration, motion games, remote control of TVs, or art installations [1, 2, 21]). Bhuiyan [1] provides a chronologically ordered overview of gesture-controlled user interfaces. Wachs [21] surveys possible application scenarios for gestural interaction but also compiles a set of basic requirements for gesture-based systems. Technically, the cited systems either make use of multi-touch surfaces [8], near-or on-body sensors [4], or contactfree/camera-based technologies [3].

Frequently, a skeletal model is used to further improve recognition [12]. Since there now are several viable options to detect gestures fairly reliably, the design and use of specific gestures has recently become an important area of research. Wobbrock [22] presented initial research into what gestures people produce naturally and identified a set of common gestures for use on multi-touch surfaces. Jokisch carried out a survey on how users discover gestures for map interaction on large multi-touch surfaces [10]. In his study, adults (20-30 years old) easily discovered simple gestures (i.e. pinch-to-zoom) but had great difficulties finding complex ones, while children (8-9 years old) had more difficulties discovering both.

A frequent example application scenario for gestural interaction is the exploration of two-dimensional maps or simple spatial data (e. g. [6, 17]. Map interaction was also a frequent example scenario to demonstrate FTIR-based surfaces [8]. Daiber [4] introduced a set of physical multi-touch interaction primitives and let user groups choose corresponding spatial interaction tasks. Recently, the manipulation of a virtual globe has also received some attention [4, 11, 19]. Boulos [11] introduced a system for gestural interaction with Google Earth and Google Street View by describing gestures for the basic functionalities. Stannus [19] reported on a comparison study, where they asked participants to perform different common navigation tasks with a virtual globe using either a mouse, a 3D-mouse or gesture control. They found that users often prefer classical input methods, i. e. a standard mouse for finding places in Google Earth, only displaying the standard imagery and no complex data. The gestural interaction was rated and commented positively in terms of naturalness. The study was of a qualitative nature, having only 10 paricipants and reported about technical problems in the prototype for the gestural interaction.

The work presented in this paper also investigates gestural interaction with a virtual globe but extends previous work by enabling interaction with the temporal dimension (to explore historical data and trends). In contrast to previous work, the prototypical application we used relied on linked spatiotemporal data on deforestation in South America. To the best of our knowledge our effort to provide exploring Linked Data with gestural interaction methods is first of its kind.

3 Exploring Data about the Brazilian Amazon Rainforest

In order to investigate the suitability of gestural interaction for the exploration of linked spatiotemporal data, we designed an interactive system aimed at enabling untrained users to explore linked data sets and visualizations on the deforestation of the Amazon region in South America. The system was exhibited at a science fair, where we gathered feedback from visitors. The physical setup of our system was based on a triangular construction with three rear projection surfaces (see Figure 1 and Figure 2).

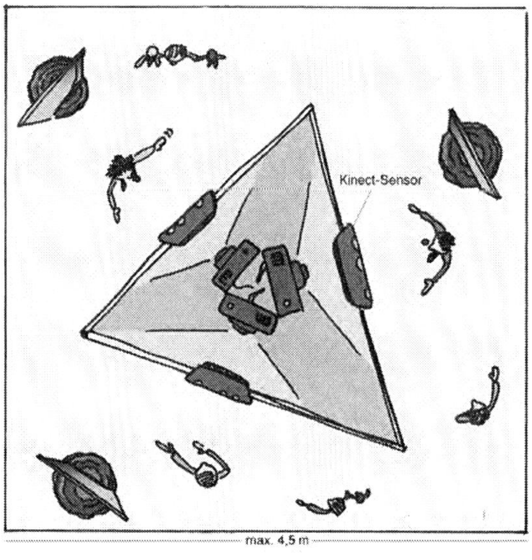

Figure 1: Physical setup of prototype: three back-projection screens arranged in a triangular configuration; each screen is controlled by a Kinect motion tracking device.

Figure 2: Gestural Interaction with the prototype on two sides providing access to different facets of the complex data se.

Each of the three screens was controlled by a dedicated gesture tracking device (i.e. a Kinect sensor). The data was rendered on the screen using a modified version of the free open source system NASA World Wind Java SDK[9] (a virtual globe system). Gestural interaction was implemented via the open source API for natural interaction OpenNI[10], a computer vision middleware NITE[11] and time series analysis [20].

The Linked Data in use represents deforestation observations from the Brazil Amazon Rainforest. Moreover, there are related social, economical and ecological data from selected Open Data sources. The social, economical and ecological dimensions were selected following the Triangle of Sustainability concept [18]. The software supports several operations to explore the data: panning—i.e. moving the map—, zooming in and zooming out, as well as moving forwards and backwards through time. The temporal operation is particularly important to support understanding of the effects and spread of the deforestation over time, and correlated phenomena like increased GDP and increased population.

The approach is completely based on free and open source software technologies and open data, making the exploration of the Brazilian Amazon Rainforest deforestation available to a broad audience in science, education, politics and society. In the next sections we introduce the open datasets and describe their visualization in NASA World Wind and the user interface with its gestural interaction.

3.1 Linked Brazilian Amazon Rainforest Data

Linked Open Data (LOD) is a set of standards and practices for exposing data on the web. LOD also allows to explicitly store the various connections in the data. We made use of our earlier contribution to open up and link the data [13] about the Brazilian Amazon Rainforest for the visualization. These statistical and spatial datasets were described and structured using mainly two vocabularies, namely the Open Time and Space Core Vocabulary (TISC)[12] and the Open Linked Amazon Vocabulary (OLA)[13] . The entire dataset originates from official sources of the Brazilian Government, namely from IBGE[14] (Brazilian Institute of Geography and Statistics) and INPE[15] (Brazilian National Institute for Space Research).

Due the recently announced policy of the Brazilian government to publish all their statistical data on the web[16] it was possible to retrieve this data and integrate it into the existing Linked Data about the Brazilian Amazon Rainforest. We realized the retrieval via a custom-made application, which connected to the IBGE's servers, downloaded the relevant data and transformed it into Linked Data to be used for the visualizations and exploration.

3.2.1 Statistical Dataset

The statistical dataset, containing a time-series from 2004 to 2009, includes the following information:

Soybeans – Soybean crops classified by hectares of planted and harvested area, value in thousand of reais (Brazilian Currency), metric tons and kg hectare. Those crops are linked to an observation and additionally linked to their respectively entries in DBPedia.

GDP – Gross Domestic Product from each municipality of Pará state.

Cattle – The total head of cattle grouped by municipality.

[9] http://worldwind.arc.nasa.gov/java
[10] http://www.openni.org
[11] http://www.primesense.com/solutions/nite-middleware
[12] http://observedchange.com/tisc/ns
[13] http://observedchange.com/amazon/ns
[14] http://www.ibge.gov.br/english/
[15] http://www.inpe.br/ingles/
[16] http://acessoainformacao.ibge.gov.br/en/

Population per municipality – Census 2000 and 2010, and population projections from 2001 to 2009.

3.2.2 Spatio-temporal Dataset

Observations of deforestation were made using remote sensing techniques, i.e. satellites. Similarly, land use data and natural and social factors of change were collected. All these data, structured and provided by INPE[5], were aggregated to grid cells of 25 km x 25 km. In Figure 3, a visualization of the dataset can be seen, where the deforestation values from cells, which spatially overlap the municipalities of Pará in Brazil, were aggregated and were plotted together with the corresponding population on the virtual globe. The spatial dataset, also containing a time-series from 2004 to 2009, has the following information:

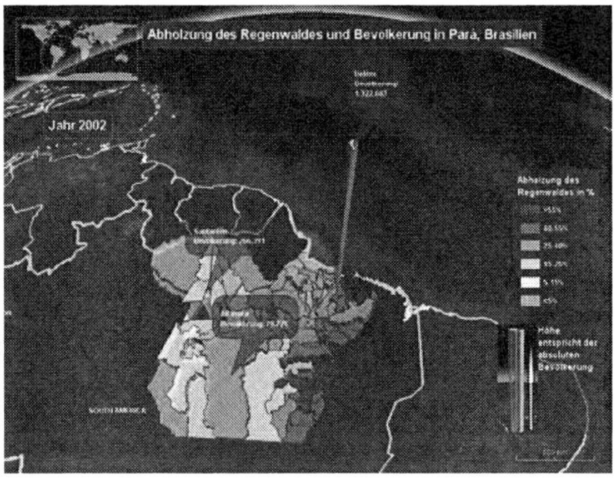

Figure 3: Visualization of the LOD Datasets on Population (height) and Deforestation (color) in NASA World Wind.

Deforestation – The observed amount of deforestation aggregated to grid cells.

Municipalities and Federal States – A complete compound of Brazilian Federal States, which belong to the Amazon area (according to the legal definition) and their municipalities, together with their geographical location and covered area.

Mesoregions – Spatial dataset that contains the mesoregions[17] area in Pará state. With this variable it was possible to group the statistical variables, originally grouped by municipality, to their corresponding mesoregion.

3.2 Visualization

The Brazilian Amazon Rainforest Dataset was visualized in NASA World Wind, divided in three different variables with the deforestation phenomena, namely GDP, Population and Head of Cattle. Each variable can be projected on the screens of the triangle and correlated with the yearly amount of deforestation, so the user can navigate year by year through the spatio-temporal datasets and observe its correlation with the deforestation phenomena. Figure 4 shows an example of the deforestation correlated with the GDP of each state of Pará in the year 2005, where the geometries height represents the GDP per person and their colour the amount of deforestation.

Figure 4: GDP (per person) and yearly deforestation in 2005.

Together with the variables and the deforestation visualization, several facts and pictures, related to each variable, were displayed in the application as annotations. These facts and pictures were programmed to be displayed according to the globe altitude, in a way that the user could have additional information to the application current view. Figure 5 shows an earth view of the Brazilian Amazon deforestation correlated with population in the year 2008. By this altitude the application was programmed to display in the left-hand side facts relevant to the current view. The facts displayed in the Figure 5 left side, written in German, say: "The pop-

[17]Mesoregions are subdivisions of Brazilian states. These subdivisions group various municipalities, based on neighboring municipalities and their common characteristics. They were created by the Brazilian Institute of Geography and Statistics for statistical purposes.

ulation of Pará represents approximately the same population of Switzerland with an area 36,75 times larger" and "Pará, population: 7.321.439, Total Deforestation: 24,13%, Population Growth: 255.920". This facts are hidden, giving place to other facts, if the user zooms the globe in or out. In the Figure 5 right-hand side, a legend is displayed to classify the geometries colours and and below it an image that explains the meaning of the geometries height, which in this case is population.

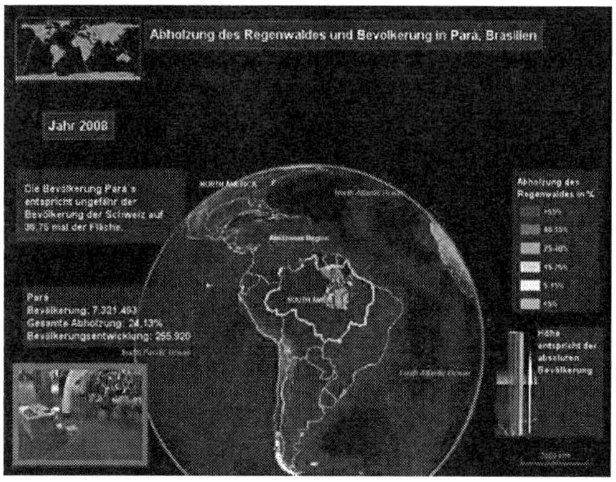

Figure 5: Earth view of the Brazilian Amazon deforestation correlated with population in the year 2008.

3.3 Gestures for Interacting with Spatiotemporal Data

Due to the importance of temporal control during the exploration of the datasets, we tried to come up with a gesture, which is easy to learn and to perform, and which also has a natural mapping to the temporal domain. Our initial approach was to use a clock metaphor, which would let the user move the hand of a virtual clock to control time by rotating their hand clockwise/counterclockwise. However, after some initial investigations we discarded this idea due to concerns about awkward and uncomfortable movements as well as difficulties in recognizing it reliably.

Eventually, we implemented three different gestures to control the spatial and temporal attributes of our map. In designing them, we tried to mimic known gestures for multi-touch enabled surfaces such as pinch-to-zoom and touch-and-drag-to-pan. While we were able to successfully apply this principle for the gestures controlling the spatial aspect, we did not find widely-used, well-known gestures for temporal control and thus designed them from scratch. The system currently supports the following gestures (cf. Fig. 6):

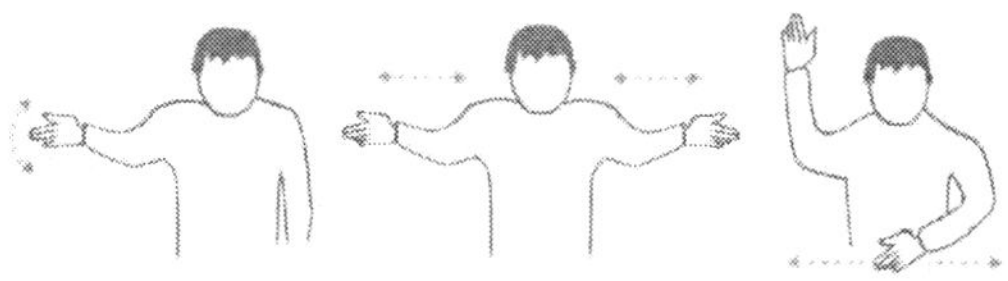

Figure 6: Complete gesture set implemented in the prototype (from left to right): one hand wipe (panning), two hand spread (zooming) and wiping with hand above head (controlling time).

One hand wipe – The first gesture is a wiping motion with one hand, which can be moved in any direction. This gesture allows the user to control which section of the map is shown, i. e. to pan the map in the direction of the hand motion. It is possible to perform this gesture with either arm, but not with both at the same time.

Two hand spread – The second gesture is used for zooming in and out of the map, i. e. to change the scale of the map. To perform this gesture, users have to move both hands and forearms simultaneously. If they move them closer together, the map is zoomed in, and if they move them further apart, it is zoomed out.

Wiping with hand above head – The third gesture controls the temporal aspect of the map. Users can carry it out by raising one hand over their head and – while keeping it above head – performing a horizontal wipe gesture with the other hand. Wiping to the left moves the timeline back by one year, wiping to the right moves it forward by one year. Visitors to the science fair could learn about the gestures by studying one of the posters we set up around the system, by observing other visitors interacting with the system, by talking to previous visitors to the exhibit, or by directly asking a member of staff continuously present near the system. The gestures can be seen in the project video[18].

4 Evaluating the Approach

In order to gain a first understanding of gestural interaction with spatiotemporal data, we carried out a

[18] `http://goo.gl/oRTFO`

questionnaire-based survey amongst visitors to the exhibit. Our goals were to get insights into how easy people found it to use and learn the gesture set, and whether it enabled them to get a deeper understanding of the complex spatiotemporal data they explored. In addition, we were interested in the fitness of the gesture set for the actions they were mapped to.

4.1 Procedure

We opportunistically asked visitors of the science fair, who had used our exhibit for several minutes (i. e. the gesture controlled exploration system for deforestation data) whether they would like to participate in our study. If they agreed, we asked them to fill out a short questionnaire. In addition, we informally observed the use of the system while noting any patterns, behavior, incidents and comments we became aware of. Participants were thanked for taking part in the study but did not receive any payments in return for their time.

4.2 Material

The questionnaire we used consisted of 28 open and closed questions in total. Participants were first asked to provide some demographic information about themselves such as their age, gender and their primary hand. We also gathered information about people's familiarity with common commercial systems that provide gesture control interfaces (e. g. Nintendo Wii, Sony Playstation Move/Eye). The second group of questions investigated task load for each gesture and were taken from the NASA TLX questionnaire [9]. We then asked participants questions relating to the use of gestures to control the spatiotemporal data. They had to rate how well gestures and specific map actions fit together, and to indicate whether they thought the gestural interaction interfered with the exploration and understanding of the linked Brazilian Amazon data being displayed. Finally, participants had to rank the three gestures overall.

4.3 Results

In total, we collected 43 responses. 28 of the participants were male and 15 female. The average age was 21.8 (mean deviation of 6.67); the youngest participant was 10, the oldest one 59. Seven of participants were left handed, 36 right handed. 19 participants stated they are familiar with gesture-based interfaces in general, while the other 24 stated the opposite. However, when asked about specific commercial systems (e. g. Nintendo Wii), it became apparent that some people had not considered these to be gesture-based systems when answering the previous question: 23 people indicated familiarity with Nintendo Wii, four with PS move, six with PS Eye Toy and eight with Microsoft Kinect. Two participants named other gesture-based systems. For this question, multiple answers per participants were allowed.

Figure 7 summarises the results for task load according to the standard NASA TLX questionnaire. The questionnaire uses a scale from zero to 20, where lower values correspond to lower work load, performance, effort or frustration, and higher values to higher work load etc. Participants rated all three gestures in a very similar way. Ratings for mental, physical and temporal demand ranged between 2.79 (std. dev. 3.23) and 4.07 (std. dev. 4.10), with the temporal control gesture generally being rated slightly worse than the other two. Performance ratings were highest for the two hand spread gesture (14.23, std. dev. 3.75), followed by the single hand wipe (13.12, std. dev. 4.2) and the wiping with hand above head gesture (12.67, std. dev. 4.6). The results in the effort category were nearly identical, whereas there were notable differences in the frustration category: here, the temporal control gesture attracted the worst rating (3.77, std. dev. 3.86), while the other two gestures were rated very similarly (single hand wipe: 3.79, std. dev. 3.51 and two hand spread: 3.93, std. dev. 3.81).

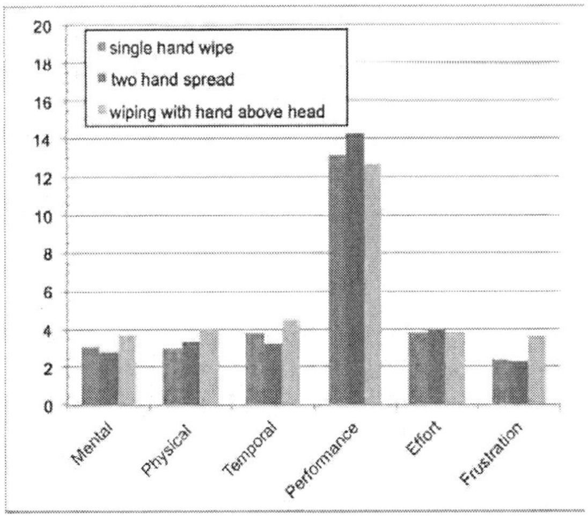

Figure 7: NASA TLX (task load) results for the three different gestures.

Results of the ranking task are summarised in Table 1. Overall, participants slightly preferred the two hand spread over the one hand wipe; the wiping with one hand above head gesture was clearly the least preferred gesture amongst the three implemented. We recorded a similar trend in terms of how well participants thought the gestures mapped to the associated actions. The wiping with hand above head gestures was rated lowest at 13.07 (std. dev. 5.14), while the two hand spread received a 14.02 rating (std. dev. 4.18) and the one hand wipe a score of 14.23 (std. dev. 3.75). The question, if the gestural interaction was rather distracting or helpful in the exploration and understanding of the linked spatiotemporal data on the Brazilian Amazon deforestation was answered with an average score of 13.92 (std. dev. 4.52). Ratings were out of 20 from zero being worst to 20 being best.

Table 1: Overall ranking of three gestures (1: best to 3: worst); the table lists number of participant, who ranked each gesture either 1, 2 or 3.

Rank	1	2	3	Average
One hand wipe	18	16	6	1.7
Two hand spread	22	12	7	1.65
Wiping with hand above head	1	12	27	2.53

4.4 Observations

In addition to the formal survey, we were also observing visitors interacting with the system: after having used our system, a number of people reported that they found the controls mentally challenging during the first few minutes, and were only able to engage with the content after that period of time. Some visitors were confused by the direction of the temporal change as they were expecting a change in the opposite direction. One participant commented in the questionnaire that our gesture is the opposite of turning the pages in a calendar. In addition, were received several positive comments from older people, who found the system very accessible, particularly if compared to their first time use of a mouse. However, there were also a number of people were unwilling to try out the system at all.

Furthermore, we noted some technical issues. The time between recognising a user and registering them as the user controlling the system increased considerably with the number of bystanders/onlookers 'seen' by the Kinect sensor. We also observed some problems in recognising children, which depending on their arm length were not detected very well.

5 Discussion

Generally speaking, the results indicate that participants were able to easily learn and use all three gestures without excessive work load, effort or frustration. Possibly due to people's lack of familiarity with the gesture used to control time, the gesture consisting of wiping with one hand above head was rated slightly worse than the other two gestures. It was particularly encouraging that participants considered the gestures helpful to engage with the complex spatiotemporal dataset visualised by the application. In addition, several older users positively commented positively on the system and its interface, indicating that gestural control could hold some potential to reach this growing user group as well.

However, since we were not able to specify specific tasks or learning goals beforehand, it is unclear whether this self-assessment reflects actual learning performance. We also did not measure actual performance or error rates during the study so that in this area, further research is required to confirm whether ease of learning and use translate to good performance as well.

Finally, we would like to mention the challenge of displaying large amounts of Linked Spatiotemporal Data on virtual globes. In our solution we chose to use only the data from Pará state since it is an interesting and representable part of the complete dataset about the Brazilian Amazon Rainforest. The data from Pará state already has very interesting combination of each chosen dimension (ecological, economical and social factors). In addition we visualized links to other countries based on the export statistics. Given these experiences we see very potential research challenges emerging for visualizing larger datasets on virtual globes.

6 Conclusions

In this paper we investigated the use of gestural interaction with complex spatiotemporal data, aggregated from Linked Open Data sources. We introduce a prototype virtual globe application, based on open source software technologies. We proposed three gestures to explore multi-dimensional deforestation data, two of which were inspired by common multitouch gestures, and one designed from ground up to control the timeline. In order to evaluate the gestures and the overall approach, we conducted a survey study at a science fair. The results indicate that the gestures were easy to learn and use (possibly after a brief initial learning phase), and were considered helpful in engaging with the complex data visualised by the application. This was true for the different age groups that participated.

In summary, our findings provide initial evidence that gestural interaction can also be successful in more complex scenarios, and thus deserves further research. One promising next step would be to conduct a study investigating naturally produced gestures (following the method pioneered by Wobbrock [22]). Additionally, modifying gestures by sensing hand and finger postures could be very useful when interacting with complex spatiotemporal data (e.g. controlling the number of years time is moved forwards or backwards, by extending the corresponding number of fingers while performing the gesture).

In terms of visualizing connections between very different phenomena —ecological, economical and social— Linked Data offered a good basis since connections could be made explicit via aggregation procedures, and via using time and space as major integrators. As a contribution the Linked Data about the Brazilian Amazon Rainforest we created and used is available also for others to use in visualizations and applications.

By interlinking aspects of Human-Computer Interaction, Linked Open Data approaches and novel visualization techniques for complex spatiotemporal data via open source software we contributed towards a Linked Open Science [14] to support transparency and openness of science.

References

[1] Bhuiyan, M., and Picking, R. Gesture-controlled user interfaces, what have we done and what's next? Tech. rep., Centre for Applied Internet Research (CAIR), Wrexham, UK, 2009.

[2] Bhuiyan, M., and Picking, R. A gesture controlled user interface for inclusive design and evaluative study of its usability. JSEA 4, 9 (2011), 513–521.

[3] Cohen, C., Beach, G., and Foulk, G. A basic hand gesture control system for pc applications. In Applied Imagery Pattern Recognition Workshop, AIPR 2001 30th (oct 2001), 74 –79.

[4] Daiber, F., Schöning, J., and Krüger, A. Whole body interaction with geospatial data. In Smart Graphics. IBM Int. Symposium on Smart Graphics, May 28-30, Salamanca, Spain, In: A. Butz, B. Fisher, and M. Christie, Eds., vol. 5531/2009, Springer (2009), 81–92.

[5] de Espindola, G. M. Spatiotemporal trends of land use change in the Brazilian Amazon. PhD thesis, National Institute for Space Research (INPE), S[2DC?]ao José dos Campos, Brazil, 2012.

[6] ESRI Inc. ArcGIS WFP API controlled by kinect. `http://www.youtube.com/watch?v=oICqUeIQ1Ko`. (accessed 02-22-2013).

[7] Fuhrmann, S., MacEachren, A., Dou, J., Wang, K., and Cox, A. Gesture and speech-based maps to support use of gis for crisis management: a user study. AutoCarto 2005 (2005).

[8] Han, J. Y. Low-cost multi-touch sensing through frustrated total internal reflection. In Proc. of UIST '05, ACM (New York, NY, USA, 2005), 115–118.

[9] Hart, S. G., and Stavenland, L. E. Development of NASA-TLX (Task Load Index): Results of empirical and theoretical research. In Human Mental Workload, P. A. Hancock and N. Meshkati, Eds. Elsevier, 1988, ch. 7, 139–183.

[10] Jokisch, M., Bartoschek, T., and Schwering, A. Usability testing of the interaction of novices with a multi-touch table in semi public space. In: HCI: Interaction Techniques and Environments, Springer (2011), 71–80.

[11] Kamel Boulos, M., Blanchard, B., Walker, C., Montero, J., Tripathy, A., and Gutierrez-Osuna, R. Web GIS in practice: a microsoft kinect natural user interface for Google Earth navigation. International Journal of Health Geographics 10, 1 (2011), 45.

[12] Kar, A. Skeletal tracking using microsoft kinect the microsoft kinect sensor. Methodology 1 (2010), 1–11.

[13] Kauppinen, T., de Espindola, G., Jones, J., Sánchez, A., Gräler, B., and Bartoschek, T. Linked Brazilian Amazon Rainforest Data. Semantic Web Journal 1 (2012).

[14] Kauppinen, T., and de Espindola, G. M. Linked Open Science— communicating, sharing and evaluating data, methods and results for executable papers. Proc. of the International Conf. on Computational Science (ICCS 2011), Procedia Computer Science 4, 0 (2011), 726–731.

[15] Malik, S., Ranjan, A., and Balakrishnan, R. Interacting with large displays from a distance with vision-tracked multi-finger gestural input. In Proceedings of the 18th annual ACM symposium on User interface software and technology, ACM (2005), 43–52.

[16] Owen Linderholm. Kinect Introduced at E3. `http://blogs.technet.com/b/microsoftblog/archive/2010/06/14/kinectintroduced-at-e3.aspx`. accessed 06-19-2012.

[17] Rauschert, I., Agrawal, P., Sharma, R., Fuhrmann, S., Brewer, I., and MacEachren, A. Designing a human-centered, multimodal gis interface to support emergency management. In Proc. 10th ACM international symposium on Advances in GIS, ACM (2002), 119–124.

[18] Serageldin, I. Promoting sustainable development -toward a new paradigm. In Valuing the Environment, I. Serageldin and A. Steer, Eds., Proceedings of the first annual international Conference on Environmentally Sustainable Development, World Bank (1995), 13–21.

[19] Stannus, S., Rolf, D., Lucieer, A., and Chinthammit, W. Gestural navigation in google earth. In Proceedings of the 23rd Australian Computer-Human Interaction Conference, OzCHI '11, ACM (New York, NY, USA, 2011), 269–272.

[20] The OpenNI organization. OpenNI home. `http://www.openni.org/`. accessed 02-22-2013.

[21] Wachs, J. P., Kölsch, M., Stern, H., and Edan, Y. Vision-based hand-gesture applications. Commun. ACM 54, 2 (Feb. 2011), 60–71.

[22] Wobbrock, J. O., Morris, M. R., and Wilson, A. D. User-defined gestures for surface computing. In Proc. of the 27th int. conf. on Human factors in computing systems, CHI '09, ACM (New York, NY, USA, 2009), 1083–1092.

[23] Wu, M., and Balakrishnan, R. Multi-finger and whole hand gestural interaction techniques for multi-user tabletop displays. In Proc. of the 16th annual ACM symposium on User interface software and technology, ACM (2003), 193–202.

AEGIS

A state-of-the art component based spatio-temporal framework for education and research

by Roberto Giachetta

Department of Software Technology and Methodology, Eötvös Loránd University, Budapest, Hungary. groberto@inf.elte.hu

Abstract

In past years, geoinformation has gained a significant role in information technology due to the spread of GPS localization, navigation systems and the publication of geographical data via Internet. The inclusion of semantic information and temporal alteration has also become increasingly important in GIS. The overwhelming amount of spatial and spatio-temporal data resulted in increased research effort on processing algorithms and efficient data management solutions.

This article presents the AEGIS framework, a currently developed spatio-temporal data management system at the Eötvös Loránd University, Faculty of Informatics (ELTE IK). This framework will serve as the future platform of GIS education and research at ELTE IK. It aims to introduce a data model for the uniform representation of raster and vector data with temporal references; to enable efficient data management using specialized indexing; and to support internal revision control management of editing operations. The framework offers a data processing engine that automatically transforms operations for distributed execution using GPGPUs, allows fast operations even with large datasets and high scalability with regard to new methods.

To demonstrate the usage of the system two prototype applications – segment-based image classification and agent-based traffic simulation – are also presented.

Keywords: Geospatial Information Systems, spatio-temporal data, indexing data structures, GPU based data processing, revision controls of data, remotely sensed image classification, agent based traffic simulation.

1 Introduction

Geographical information systems (GIS) have undergone a spectacular development in the past years. Beside traditional areas of GIS applications a rapid development has taken place in the world of navigation systems as well. Google Maps and NASA World Wind, together with their Application Programming Interface (API), are common tools in the global handling of spatial data. The world of open source software has also evolved a lot. There is a rising need for professionals whose practice cover both information technology and geography. This paradigm shift has to be taken into account both by professionals and by academic people.

At the Eötvös Loránd University, Faculty of Informatics (ELTE IK) the informal association, called Creative University GIS Workshop (TEAM) deals with several related research topics, e.g. Intelligent Raster image Interpretation System (IRIS), University Digital Map Library (EDIT), Virtual Globes Museum (VGM) and segment-based analysis of remote sensing images. An important collaboration takes place both in education and research with the Institute of Geodesy, Cartography and Remote Sensing (FÖMI). This governmental institution is responsible for the research, development and application of remote sensing in Hungary, mainly in the areas of agriculture and environmental protection.

Based upon the experience gained with research and education, the plan of a standalone geographic framework, called AEGIS has been outlined. It is designed for broad functionality, efficiency, and allowing students to skip the learning curve and concentrate on their main tasks in lab projects and thesis works without the need of building up auxiliary functionality from scratch. The system is based on standards of the Open Geospatial Consortium (OGC – `http://www.opengeospatial.org`), and is not intended to be a competitor of current industrial GIS solutions or GIS toolkits such as GeoTools (`http://www.geotools.org`), but rather to be a prototype and sandbox for experimental research and education.

Students are offered step-by-step learning of standards, algorithms and solutions, easy to understand, documented source code, and well structured components that can be used for many types of GIS projects, including thesis works. Researchers may also find many possibilities in this flexible, yet stable environment to test results and compare them with existing solutions. Even the framework itself has several experimental features that are research topics, including multi-level spatio-temporal indexing solutions, revision control of spatial data and automated

GPU based data processing.

The AEGIS concept has been introduced in Giachetta et al. (2012), and is currently under development. Some features are only outlined in the paper to be implemented in the future. The framework is to be released as open-source project in the future.

The rest of the paper is arranged as follows. In Section 2 we will outline the concept and architecture of our system. Section 3 explains details of the AEGIS data management model. Section 4 presents the data processing capabilities and solutions. Section 5 describes the first applications of the system, namely segment based image classification and agent-based traffic simulation. Section 6 concludes the paper.

2 System architecture

AEGIS has component based architecture in order to be as adaptable and customizable as possible. Several class libraries are stacked together to build up system functionality and relevant libraries can be chosen for any application ranging from desktop editors through mobile viewers to server services. These libraries also provide API functionality for building custom applications. The AEGIS framework stack contains the following components (see Figure 1).

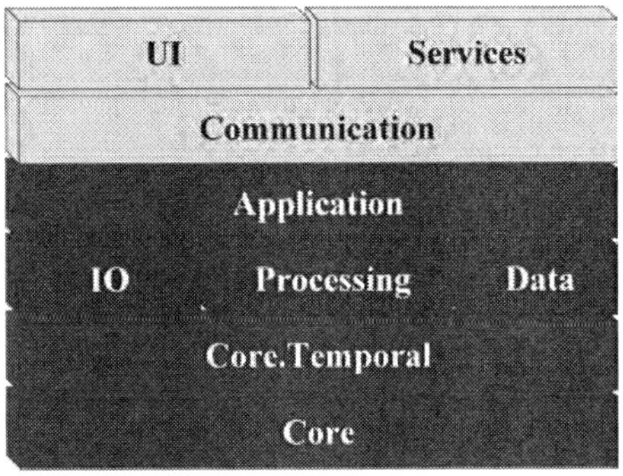

Figure 1: AEGIS framework stack.

- *Core* is the heart of the framework containing the OGC and ISO compliant 2D/3D spatial data model and reference model including several utility classes (e.g. indexing structures). This component is designed not to contain any temporal references as not all GIS applications require it. However base interfaces are provided for time management.

- *Core.Temporal* is the extension of the data model with temporal properties, reference model and spatio-temporal indexing structures supporting multiple time measurement solutions. Our solution focuses less on moving objects and trajectory and more on modeling stationary spatial features with limited temporal validity, for instance, road sections closed due to constructions and remote sensing image series of crop growth. Details of the model can be read in Section 3.

- *IO* contains the import/export possibilities for GIS file formats and network services. Several standard spatial formats are supported. The module allows the consumption of OGC web services including WMS and WFS. Support for further services is planned.

- *Data* component is responsible for database management. All spatio-temporal data imported into the system can be stored in a database. AEGIS is designed to handle spatial and spatio-temporal information (including indexing, revision control, etc.) entirely within the system, hence no spatial features of the database system are required. It is clear that this solutions lacks the effectiveness and potential offered by most database management systems. However it provides researches a stable high level environment to work in. Based on our earlier experience, MongoDB (`http://www.mongodb.org`) has been chosen as the primary database backend.

- *Processing* offers spatio-temporal vector/raster processing operations. For both the Processing and the IO components functions are defined using a meta descriptor system to support runtime management and extension. Processing services also offer an internal programming scheme based on expression trees. By using this scheme the processing instructions can easily be transformed to low level GPU instructions and automatically distributed. An internal processing engine is responsible for this purpose. Details of processing can be read in Section 4.

- *Application* layer is responsible for user authentification, editing, processing and import/export management including the runtime handling of available IO and Processing functions. During feature editing changes are overseen and logged to enable revision control of the data. The layer also contains a scripting engine using a dedicated high level domain specific language to enable the creation of custom processing operations and batch processing of data even from multiple sources and formats.

- *Communication* layer contains services for internal communication between application frontend (User Interface and Services) and backend (Application layer). The primary objective of the layer is to enable the separation of these two layers and the creation of thin clients that can execute operations on remote servers. Communication uses a service oriented architecture with encryption and user authentification.
- *Services* is responsible for providing web services, mainly in OGC standard formats (WMS, WPS, etc.). This server side module is not to be confused with the client side IO component which offers OGC service clients to enable import of data into the system.
- *UI* provides graphical user interface components and 2D/3D data visualization methods for multiple platforms including desktop and mobile environments. There are also plans for extending visualization using augmented reality on mobile devices.

Using the component based design, the structure of the system can be configured with each installment. For instance, a thin client installation of AEGIS may only require Core, UI and Communication components as data storage, processing, import and export are performed by the server which is connected to the client via Communication layer. In contrast the server does not require UI and IO modules, but should include Services to enable access to data for other applications. A thick client may contain all components except Services and can support local or remote execution of processes. Figure 2 shows this layout possibility.

The implementation is carried out using the Microsoft .NET Framework, because of the wide possibilities and the simple usage of this development platform, the strong .NET education carried out at ELTE and also the success of .NET based GIS software products such as DotSpatial (`http://dotspatial.codeplex.com`) and SharpMap (`http://sharpmap.codeplex.com`).

In the current phase of development the higher layers (Application, Communication, Services and UI) are only partially complete. Additional research, design and testing will follow before these components are implemented.

3 Spatio-temporal data management

Modeling and indexing of spatio-temporal objects has been a frequent topic among researchers since the 90s, and many data models have been introduced (Abraham & Roddick 1999, Nguyen-Dinh et al. 2010). However for spatio-temporal data neither common solutions have arisen, nor standards have been developed so far. In contrast, spatial data has well known standards maintained by OGC.

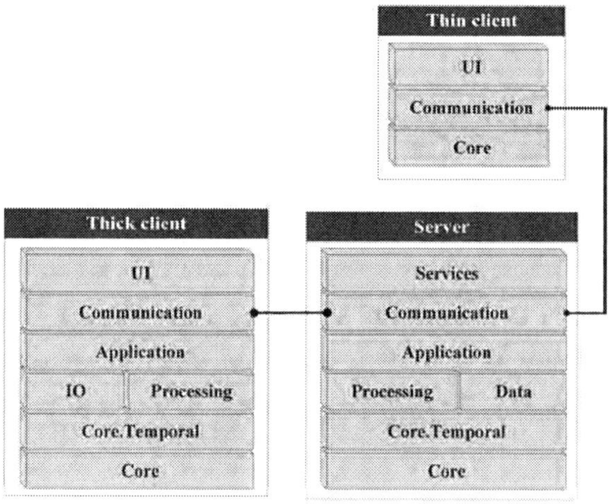

Figure 2: Example layout of AEGIS components with thin and thick clients.

As explained in Section 2 the AEGIS data model is implemented in two components.

- *Core* holds the spatial model including raster and vector data representation, spatial indexing structures and reference system management. Basic interfaces are defined for temporal relations but are not implemented in the component. The usage of Core components is recommended in applications where no temporal properties are needed. For all spatial vector features the representation is a direct mapping of the OGC Simple Feature Access standard (Herring, J. R., ed. 2011). However the standard has been extended with several new types, including raster support. Detailed description of this model follows in Section 3.1. The reference model is an implementation of the OGC Spatial Referencing by Coordinates specification (Cooper, P., ed. 2010) using EPSG guidelines (`http://www.epsg.org/guides/`).

- *Core.Temporal* extends the core model with temporal properties and references. Hence base classes and interfaces are imported from the Core package, the spatio-temporal model can function as strictly spatial if needed.

Besides adding temporal properties to features spatio-temporal indexing is introduced. Temporal references are managed with multiple temporal reference systems (e.g. Clocks, Calendars). The reference model is a partial implementation of *ISO 19108:2002 Geographic information – Temporal schema* (2002).

3.1 Unified vector and raster data representation

In the OGC SFA data model the central item is the *Geometry* class, which has several specialized versions (including collections) in object inheritance taxonomy. Geometry defines the interface for spatial properties and operations among any spatial objects. The model focuses on two and three dimensional vector data without any temporal references. In our implementation of this standard, multiple extensions have been introduced to enable flexible and effective implementation of features (see Figure 3).

Vector and raster data are required to be handled in the same manner so that operations work uniformly on them. In current geospatial systems the representation and operations of raster and vector data are usually independent of each other. In our system, raster data is introduced as descendand of Geometry. All spatial operations, which are defined on vector objects (e.g. intersection, translation, projection etc.), can also be performed on raster data (for example the ability to intersect an orthophoto with a land parcel polygon). By using the meta descriptor system restrictions can be applied to any processing operation to prevent raster functions (e.g., intensity transformations, image filtering) working on vector data.

For raster data the *Raster* and *RasterBand* classes are introduced. RasterBand is a descendant of Rectangle and contains one band of a raster image, while Raster is a collection of multiple bands of an image. These classes are enhanced with operations and properties related to raster imagery. Images can be stored in several radiometric resolutions (from 8 bit to 64 bit for every band), may contain pyramid layers, and image masks (to indicate actual image pixels).

Vector geometries can be transformed to graph representation using the *GeometryGraph* class that contains points as edges and paths as vertices. This representation enables merging multiple geometries to form a single network. Equivalently the *RasterGraph* class is the graph representation of raster data where vertices contain intensity informations of pixels. Graphs may include a multi-level structure, where a single high level vertex contains multiple low level vertices. This is useful for example, managing road hierarchies or raster image segments.

New collections, called maps have been introduced primary for efficiently managing large amount of objects. Map classes are build up using R-tree based indexing structures for fast spatio-temporal query of objects, as described in Section 3.3.

All geometries have been extended by the possibility of storing the time instant or time interval in which they are considered to be valid. Generally, items in a collection do not need to exist in the same time interval, but restrictions can be made to force temporal equality among all items of the collection. The reference system of these temporal geometries is usually a compound reference system containing a temporal system to manage calendar or clock aspects.

All features may contain any number of descriptive data (metadata) stored as key/value pairs. All geometries within a collection require to have the same metadata schema. The metadata is also time-dependent and is only valid when the feature is valid, but values can change multiple times within the validity. These changes may or may not depend on the change of the feature's spatial properties.

It must be noted that in the current phase of development not all operations work as efficiently as other implementations of the SFA standard (e.g., GeoTools, DotSpatial) as clarity, readability and stucture of the source code is among top priorities, but steady progress is expected also in this factor.

3.2 Data storage

The data model has been specifically designed to handle all spatial and temporal properties of the data at system level without requiring any spatial support from the database system. Hence the database backend has been chosen based on previous experimental results on performance of queries and modifications (Máriáas et al. 2012). Hence, currently MongoDB serves this purpose. In the future support is planned for PostGIS and other database systems as well.

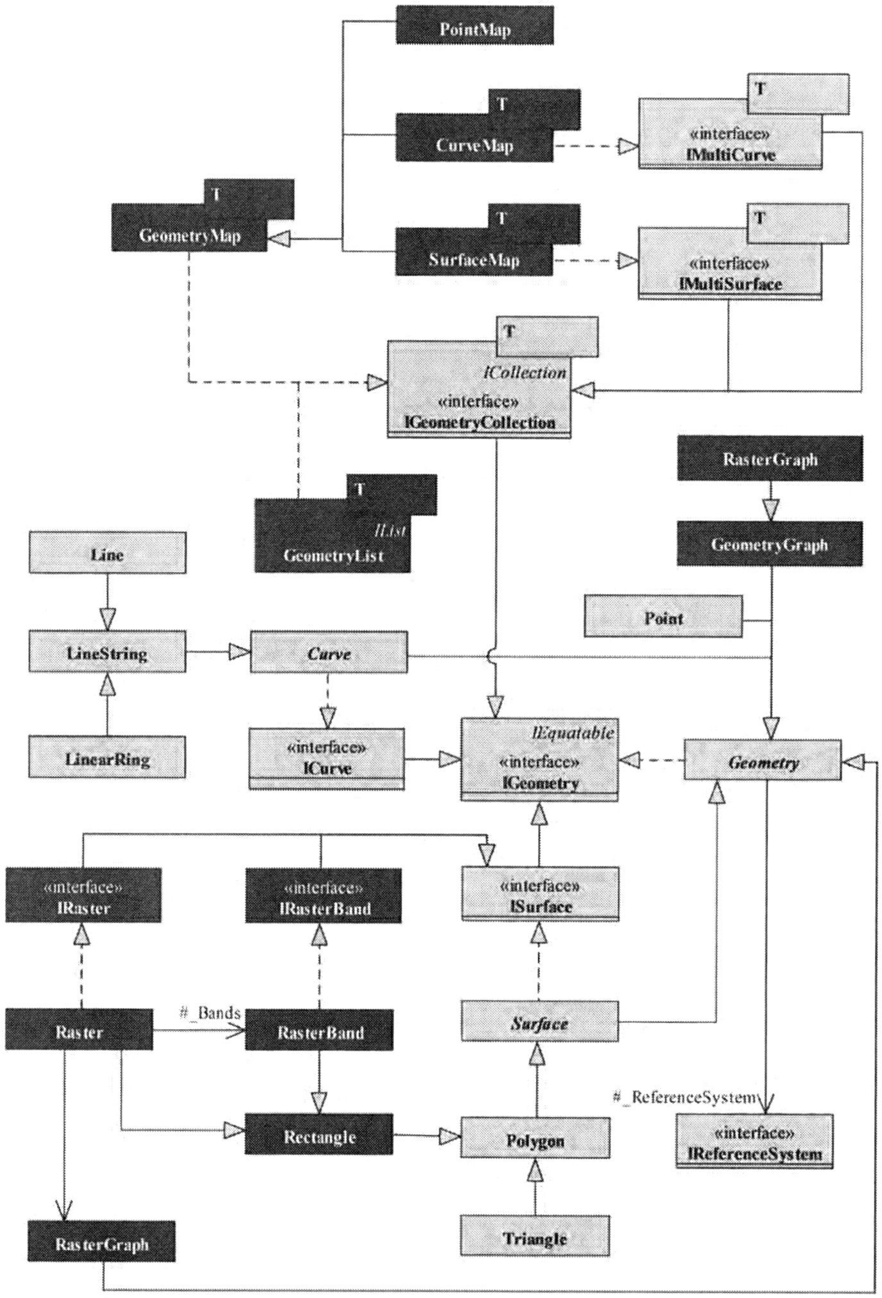

Figure 3: The data model based on Simple Feature Access.

MongoDB is a document-oriented database management system (DODBMS), which enables the schema-less storage of hierarchical data (Padhy et al. 2011). Due to the large variety of spatio-temporal and descriptive data it has proven to be more efficient in data querying than SQL based systems. MongoDB has only minimal spatial support for 2D geometry, but the structure of the stored items can be easily customized, even to allow the storage of different types of geometries in a single collection. Thus, no spatial operations function for these geometries in the DODBMS, but still, simple spatial and temporal queries (such as bounding box based selection) are usable.

For storing raster data the internal *GridFS* convention is used that is specifically designed for storing large amount of binary information linked to the document. Again it can be noted that no image operations function directly at database level.

It can be clearly seen that out solutions lacks the speed and potential of spatial database operations, but has multiple gains. The AEGIS data model can be easily mapped to the document-based structure of MongoDB, even raster data by using GridFS. Extending the database management system with new data models and indexing solutions would be far more complex than building the required infrastructure at system level, so research in these areas can progress quicker, whereas the results can be adopted into database systems in the future.

3.3 Spatio-temporal indexing of geometries and metadata

As described in Section 3.1, there are multiple temporal aspects treated in the data model.

- Features have limited temporal validity, a time interval in which they are considered to be valid. This interval may be the same for the entire collection, but can be different for each object.

- Features may change spatial properties during their validity. In the case of moving objects, the entire geometry is relocated, but is not changed. However our model enables any change in the geometry itself, even a single point may alter within a polygon leaving the rest in place.

- Metadata may change within the validity. Metadata keys and values can change independently from each other and from spatial alteration. The values indentified by the same key but with different timestamp can be independent or in relation to each other. For instance the average travel speed of a road section for a given timestamp can be expressed as a factor of the speed limit.

To support this broad spectrum of temporal management, indexing has been introduced on multiple levels. For spatio-temporal indexing *3D/4D R-trees* (Kolovson & Stonebraker 1991) and *Multiversion 3D/4D R-trees* (MV3R-trees) are used (Tao & Papadias 2001). In both cases, time is the last (third or fourth) dimension. The former is efficient for timestamp based queries, the letter has good performance in both timestamp and interval queries. Both trees have spatial and temporal bounding boxes, and use multiple heuristics to enhance the performance of tree updates. For temporal indexing of metadata, *Multiversion B-trees* (MVB-trees) are included (Kolovson & Stonebraker 1991).

Using these structures the following multi-level indexing solution was implemented, as illustrated in Figure 4).

- Map collections contain MV3R-trees to support both interval and timestamp queries.

- Features contain 3D/4D R-trees to access spatial properties at the specified timestamp.

- Metadata collections of the features use MVB-tree to query values at the specified timestamp.

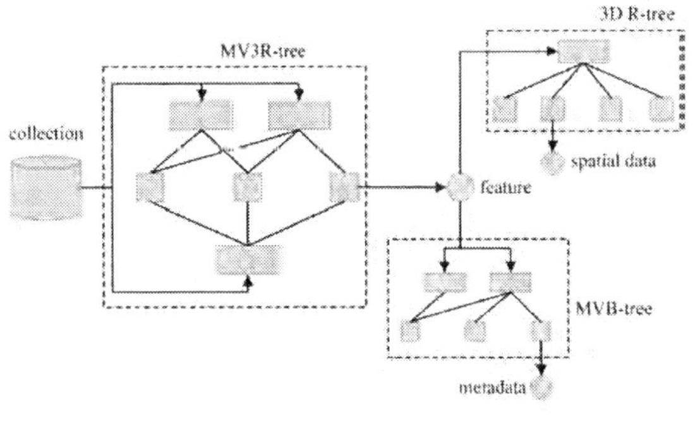

Figure 4: The multi-level indexing solution.

In certain situations some features can belong to multiple collections, and have different temporal properties in these collections. For example, a road section can belong to a collection containing daily traffic information (where average travel speeds are calculated for every hour), and also to a collection with yearly traffic information (where average travel speeds are calculated for every day). Different metadata values are need to be calculated in the two collections. In the simplest solution this would require the duplication of the feature.

To prevent this duplication, additional B+-trees have been included in collections and graphs. This can be seen in Figure 5. The solution is specifically designed for situations where relations can be established between the altering metadata. The tree contains timestamps as keys, and for each timestamp, a collection of temporal variables. These variables contain descriptive information that changes in time, and is not stored in the feature. Variables are formed by (key, modificiation, usage) triplets. These triplets define metadata variability for any descriptive value of the geometry. Key refers to property name, usage defines application of the modification, such as override, add, multiply, etc.

For instance, the speed limit may be stored as metadata in the feature, whilst the tree contains the average travel speed for a given hour as a factor of the speed limit.

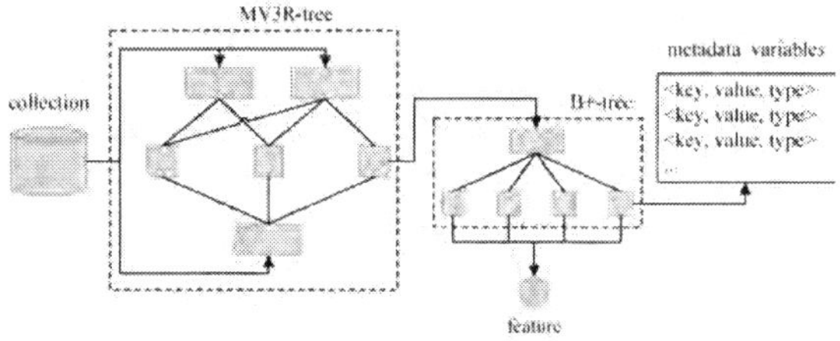

Figure 5: The MV3R-tree extended with B+-trees.

When querying the speed for a certain timestamp, the speed limit is multiplied by the factor providing the temporal variability.

The metadata variability is only reachable through the collection, and different collections provide different alterations of the metadata values. However this solution is only effective when the amount of changes and therefore the space required by the B+-trees is less than the space required for the feature duplication. Since collections may contain other collections, this effectiveness can be raised by grouping features into inner collections, where the value change factors are the same.

3.4 Revision control of data

Revision control is an essential part of modern application development. Hence it is also usable in any field where the modifications of documents are needed to be tracked. In this regard spatio-temporal features may be considered as documents where editing operation cause changes in either spatial, temporal or metadata properties of the feature. In vector representation editing operations are restricted on spatial and temporal geometry properties but in raster representation they may also be changes involving pixel data. Only a handful of studies deal with revision control of images. Modern revision control solutions offer only limited possibilities for managing binary data.

Our solution, based on Chen et al. (2011), is integrated as a part of the framework. The core idea is the storage of image operations using a directed acyclic graph (DAG) where each vertex contains a performed editing operation. This representation is suitable for nonlinear revision control and does not require the storing of the entire modified image.

In the AEGIS system, all editing operations on geometries are executed as either vector based modifications (including all raster image changes) or as simple variable changes (such as metadata modification). Therefore it is fully compliant with the graph based storage. The method is also applicable for editing vector features. However in our system a graph is maintained for the entire editing lifetime of a feature, not just between two revisions. Therefore all previous revisions are obtainable and the framework only needs to store the original raster and the head revision of each branch (if needed). Revisions may also be tagged. A tagged revision stores the entire geometry, so it can be retrieved without any need of reapplying editing operations.

The revision control process is fully automated and reacts to all changes signaled by the feature that reach the Application level.

4 Flexible data processing

Data processing, algorithmic capabilities and supported import/export formats are the key questions in most applications and services. Not only the number and purpose of processing algorithms should be taken into account but also their performance is crucial. All these aspects were taken into account during the design of AEGIS with the inclusion of dynamic extendibility and GPGPU based distributed execution.

Dynamic extension is supported by using reflection capabilities of the .NET framework and the meta descriptor service, that is used by the Processing and IO packages. Algorithms may be implemented in any .NET language using the AEGIS API, which offers many base classes for file or remote service access and algorithm description. These compiled class libraries can be simply added to the system dictionaries to be automatically recognized and usable through UI commands and the scripting engine.

The scripting engine offers an easy to learn high level procedural language, currently named *AEGIScript*. It is possible to create batch scripts on any import/export and geometry processing operation; to combine simple algorithms to form composite, reusable operations and to execute algorithms sequentially and in parallel.

The efficient execution of processing algorithms is another issue. It is widely known that intermediate language programs executing on virtual machines cannot match the speed of native applications. Relying strictly on .NET code in algorithm execution would result to poor performance. Hence algorithms can be written as expressions, which are dynamically compiled and transformed to either .NET intermediate language (IL) or OpenCL language codes. In the latter both CPU and GPU based executions are supported to enhance processing speeds. Thus the programmer may write the algorithm in simple .NET code, which can be executed as low level code on parallel GPGPU architecture. The complete structure of the processing environment can be seen in Figure 6.

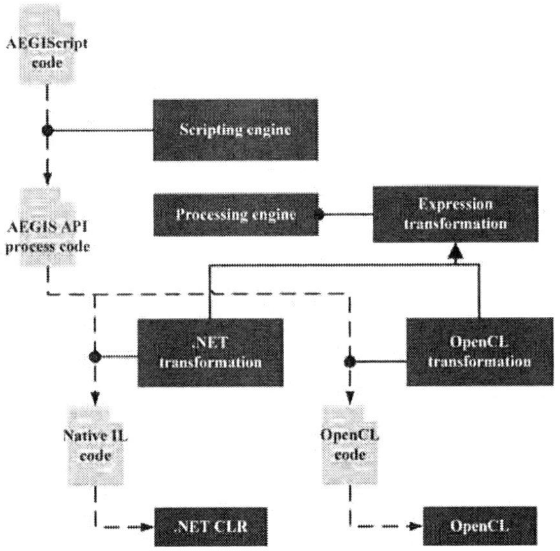

Figure 6: Control flow of the processing services.

This compilation is performed by the processing engine. The solution is effective in handling of large raster and vector data. However input data must be first converted to the specified simple format of the execution environment (e.g., simple one dimensional arrays for OpenCL) and back after the operation. The transformation is automated, but can be time-consuming. After transformation, algorithms can work either in place or out place on the data (as defined by the meta descriptor of the process).

The processing engine is also extendible with new execution environments including other GPU based models (such as CUDA) or other platforms (e.g., database system). It must be noted that in the current phase of development not all processing operations are supported by the transformation engine, these are executed as standard .NET code.

5 Applications

The first application, based on AEGIS architecture, is a simplified agent-based suburban traffic simulation. The simulation calculates and displays traffic information, mainly congestion levels for every time of the day within a year. The main goal of the application is the testing of the implemented indexing solutions with vector data. The application has been developed using the Core, Core.Temporal and IO packages of the AEGIS stack; separate operations and display environment are built on top.

The second application is a segment based classification library for remote sensing images. The algorithms have been previously implemented in both C++ and C#. Hence, the goal of the application is the comparision of performance and efficiency of the Processing component and the OpenCL transformation.

Both applications are under development, therefore results are preliminary. Definitive results will be published in the future.

5.1 Agent based traffic simulation

In this simulation a number of independent agents drive on the road nework during the day. The agents have multiple destination addresses to drive to. Two main destinations are workplace and home; but multiple random targets (like shopping centers or restaurants) may also occur. Agents travel to all locations by car. Locations are built up using metadata of building objects in the city map. Agents present in this simulation are primitive; they use simple distribution-based algorithms to make decisions. For example, an agent may have a working time from 8 o'clock in the morning to 6 in the afternoon, and can go to multiple stores or other locations after work.

Figure 7 illustrates the visualization of the traffic simulation on the map of Budapest, Hungary. In Figure 7a agents traveling to destination are shown in red, agents travelling home in green. Figure 7b displays the traffic congestion levels (yellow for small congestion, red for large congestion) for 9:00 AM of a specified day.

Figure 7: Visualization of the agent based traffic simulation: (a) Running agents; (b) Traffic congestion levels.

Agents plan their routes using *A*-algorithm* working on the *GeometryGraph* representation of the road network. Routing algorithm calculates the journey based on road travel time, which is available as metadata. The calculation is performed beforehand, based on historical traffic information combined from yearly and daily data including the last results prior to calculation. During travel the actual driving speeds are calculated based on the number of agents and the loadability of the road section. The traffic data is recalculated every 15 minutes. Due to traffic jams occuring, the agents speed can become too low or the travel time can become too high causing the agent to perform one of multiple actions, including recalculation of the entire route avoiding nearby road sections, recalculation of partial route, or even altering the destination address.

To manage the traffic data, only a single copy of every road section is used for all calculations by multiple graphs in combination with B+-trees for metadata variability storage. This solution can be seen in Figure 8. Data is summarized and optimized (including rebuilding of B+-trees) after each day. Measurements show that although optimization of the collections is quite time consuming, the query of values is only 15-20% less efficient than using multiple instances of the road sections, causing 5-8% performance loss of the entire simulation, with the benefit of using only 55-60% of storage space compared to multiple instances.

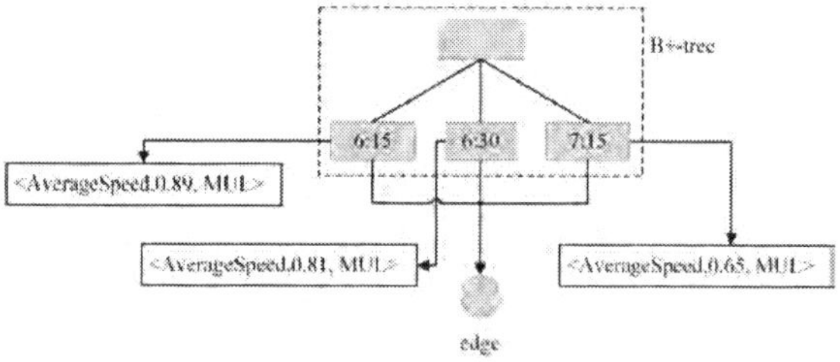

Figure 8: Temporal metadata indexing used by the traffic simulation.

The application also served for inspecting the efficiency of the MV3R-tree by blocking road sections and so limiting their temporal validity, but no significant performance differences were measured in case of acceptable amount of changes.

5.2 Segment based image classification

To achieve the proper detection of features in mapping, image objects, that fit land cover objects, are needed to be delineated and they must be classified into predefined classes. However, traditional pixel-based classification methods completely disregard spatial relations.

In recent years, scientists of ELTE and F ÖMI have been jointly investigating efficient segmentation methods of remote sensing images. Up to now the implementation and investigation of several algorithms have taken place, including the analysis of parameterization and accuracy (Dezső, Giachetta, László & Fekete 2012).

For testing and comparison purposes, several dedicated applications and libraries have been developed and some open-source and proprietary products have been used including eCognition (Dezső, Giachetta, László & Fekete 2012). However these solutions lack of flexibility and customization possibilities which are needed for accurate results. As a result implementation of several segmentation methods and segmentation based classification algorithms have begun recently in the AEGIS framework.

These approaches include both merge-based and cut-based methods, namely *sequential linking*, *best merge* and *minimum mean cut*. All algorithms work on the graph form of the image, represented by the RasterGraph class, that can store either pixel or segment information in vertices. These graphs are easily convertible to arrays that can be handled by OpenCL. By defining the steps of the algorithm through expressions, the entire process can run on GPU after the transformation of the graph with the same results as in previous implementations.

Initial results are promising regarding execution time. Using OpenCL based processing, the performance increase is up to 70-75% compared to standard .NET implementation and 30-40% compared to previous C++ implementation. However, further investigation and optimization will take place before trustworthy results can be presented.

6 Conclusion and future work

In the previous sections the concept of the AEGIS spatio-temporal framework and the goals of the authors' research have been introduced. Although development is still in early stage and several aims are still in planning phase, some results have already been reached.

The system is based on the spatio-temporal data model as described in Section 3, which uses complex data structures and multi-level indexing with temporal variability to enable more efficient management of data. Processing services allow flexible extensions and dynamic transformation of algorithms to GPGPU execution for improved performance. Processing capabilities are extended by the inclusion of batch scripting.

Further research includes the testing of other indexing structures and representation possibilities, and their performance measurement using more applications; the enhancement of algorithm transformation for wider support on different operation types and the implementation of higher levels in the AEGIS infrastructure, namely UI, Communication and Services.

References

Abraham, T. & Roddick, J. F. (1999), 'Survey of Spatio-Temporal Databases', *GeoInformatica* 3(1), 61–99.

Chen, H.-T., Wei, L.-Y. & Chang, C.-F. (2011), 'Nonlinear revision control for images', *ACM Trans. Graph.* 30(4), 105:1–105:10.

Cooper, P., ed. (2010), *The OpenGIS Abstract Specification -Topic 2: Spatial referencing by coordinates, version 4.0*, Open Geospatial Consortium.

Dezső, B., Fekete, I., Gera, D. A., Giachetta, R. & Laszló, I. (2012), 'Object-based image analysis in remote sensing applications using various segmentation techniques', *Annales Universitatis Scientiarum Budapestinensis de Rolando Eotvos Nominatae Sectio Computatorica* 37, 103–120.

Dezső, B., Giachetta, R., László, I. & Fekete, I. (2012), Experimental study on graph-based image segmentation methods in the classification of satellite images, *in* R. Vaughan, ed., 'EARSel eProceedings', Vol. 11, pp. 12–24.

Giachetta, R., László, I. & Bálint, C. L. (2012), Towards the new open source GIS platform AEGIS, *in* A. Degbelo, J. Brink, C. Stasch, M. Chipofya, T. Gerkensmeyer, M. I. Humayun, J. Wang, K. Broelemann, D. Wang, M. Eppe & J. H. Lee, eds, 'GI Zeitgeist 2012 -Proceedings of the Young Researchers Forum on Geographic Information Science', Heidelberg: Akademische Verlagsanstalt, pp. 11–22.

Herring, J. R., ed. (2011), *OpenGIS Implementation Standard for Geographic Information: Simple Feature Access -Common Architecture, version 1.2.1*, Open Geospatial Consortium. URL: http://www.opengeospatial.org/standards/sfa

ISO 19108:2002 Geographic information – Temporal schema (2002), International Organization for Standardization, Geneva, Switzerland. URL: http://www.isotc211.org/Outreach/ISO_TC_211_StandardsGuide.pdf

Kolovson, C. P. & Stonebraker, M. (1991), 'Segment indexes: dynamic indexing techniques for multi-dimensional interval data', *SIGMOD Rec.* 20(2), 138–147.

Máriás, Z., Nikovits, T., Takács, T. & Giachetta, R. (2012), 'A study of storing large amount of inhomogeneous data in workflow management systems', *Annales Universitatis Scientiarum Budapestinensis de Rolando Eotvos Nominatae Sectio Computatorica* 37, 275–292.

Nguyen-Dinh, L.-V., Aref, W. G. & Mokbel, M. F. (2010), 'Spatio-Temporal Access Methods: Part 2 (2003 -2010)', *IEEE Data Eng. Bull.* 33(2), 46–55.

Padhy, R. P., Patra, M. R. & Satapathy, S. C. (2011), 'RDBMS to NoSQL: Reviewing Some Next-Generation Non-Relational Database's', *International Journal of Advanced Engineering Sciences and Technology* 11(1), 15–30.

Tao, Y. & Papadias, D. (2001), MV3R-Tree: A Spatio-Temporal Access Method for Timestamp and Interval Queries, *in* 'Proceedings of the 27th International Conference on Very Large Data Bases', VLDB '01, Morgan Kaufmann Publishers Inc., San Francisco, CA, USA, pp. 431–440.

A new GIS toolbox for integrating massive heterogeneous GIS data for land use change analysis

by Birgit Laggner and Natascha Orthen

Thünen-Institute of Rural Studies, Braunschweig, Germany. `birgit.laggner@ti.bund.de`

Abstract

Agricultural land use in Germany and related impacts on the environment and the use of natural resources are key research topics at the Thünen-Institute of Rural Studies. As spatial context is essential for the analysis of causal connections, GIS data regarding all necessary information was gathered during different research projects and prepared for processing in a database. In particular, the Integrated Administration and Control System, which was available for certain project purposes for several Federal Laender and years, serves as a very detailed data source for agricultural land use. We use different Open Source GIS software like PostgreSQL/PostGIS, GRASS and QuantumGIS for geoprocessing, supplemented with the proprietary ESRI product ArcGIS. After introducing the used input data and the general processing approach, this paper presents a selection of geoprocessing routines for which Open Source GIS software was used. As an exemplary 'use case' for the conclusions from the consecutive statistical analysis, we summarize impacts of increased biogas production on agricultural land use change highlighting the trend in biogas maize cultivation and the conversion of permanent grassland to agricultural cropland.

Keywords: PostGIS, GRASS, geoprocessing, multiple data, Open Source, pre-processing of data, environmental impacts, biogas maize, permanent grassland, Germany.

1. Introduction

Agricultural land use and land use changes are related to environmental impacts like groundwater pollution, greenhouse gas emissions or loss of biodiversity (Schramek et al. 2012). These spark political efforts which result in legal acts (e.g. Nitrates Directive, FFH Directive). Agricultural subsidies or legal restrictions add to the driving forces influencing the usage of agricultural land, while land use must suit the natural conditions in order to optimize farm income.

The Thünen-Institute of Rural Studies analyzes land use change in agricultural areas of Germany focusing on environmental impacts and the effects of legal regulation with research projects like "Evaluation of the Common Agricultural Policy from a nature conservation point of view" (2008-2009) (see Nitsch et al. 2012) and "Analysis of land use change and development of methods for identification and quantification of measures for greenhouse gas abatement in the agricultural sector (parts I and II)" (2009-2013). A broad variety of GIS data is gathered in order to build a database for statistical analysis. The data contains official national and federal digital maps regarding land use, natural conditions and protected areas.

The central objective described in this article relates to the development of a GIS toolbox enabling the processing and analysis of massive and heterogeneous geodata useful for the statistical analysis of land use change patterns. The GIS toolbox subsumes all necessary processing steps for data preparation and for the integration of all input datasets into a combined dataset preserving relevant information on every site.

We base the development of the GIS toolbox on several conceptual considerations, whose applicabilities are to be proven as one result of the development process. Decisions with respect to the conceptual approach of the GIS toolbox include the choice of the main data format and the software environments as well as the definition of the necessary processing steps.

A conceptual precondition for the development of the GIS toolbox is the decision to chose a vector polygon approach as the main data format. This decision is motivated by the aim to preserve as much information as possible and necessary from the input data. Mostly all following conceptual decisions depend on this precondition. The related research challenges concern the feasibility of this approach within the given circumstances as well as the evaluation of its expandability and appropriateness. We are aware of the uncertainties caused by the heterogeneous resolutions of the input datasets, which have to be considered at the discussion of the results of the statistical analyses.

Another conceptual challenge relates to the applicability of a modular design for complex processing steps against the background of the data heterogeneity. While a modular structure would reduce the effort of recurring tasks (addition of data updates, processing of similar datasets for different Federal Laender), even data presumed to be very similar differ in structure and content between different dates or origins.

The necessity to find a balance between rivaling requirements when it comes to software selection represents the next conceptual challenge. First, the chosen software has to fit into the existent infrastructure in order to allow easy usage and to benefit from existing abilities. Second, the efficiency of different software solutions at handling specific tasks on massive geodata has to be considered. Third, the number of interfaces has to be minimized. In this context, it was important to evaluate the use of Open Source software as an alternative to the proprietary software ArcGIS for the intended GIS toolbox.

Challenges with respect to data heterogeneity include handling of different data formats (vector and raster data, not georeferenced data) and spatial references as well as the extraction of relevant data from differently structured sources. All datasets contain a variety of attribute information while the quality of their documentation is varying. Additionally, some input datasets consist of several files in a database-like structure. Concerning the geoprocessing aspect, the attribute-related heterogeneity mostly affects the elimination of overlapping polygons. Further geoprocessing issues derive from sliver and invalid polygons.

The analysis of the expansion of biogas production based on renewable resources from agriculture, an example for land use change induced by political actions, illustrates the use of the GIS toolbox. Superficially, the substitution of fossil energy resources with renewable bioenergy seems like a good way to reduce greenhouse gas emissions. Though, the cultivation of bioenergy crops competes with food production for the same agricultural areas and may increase environmental impacts with respect to, e.g., biodiversity and water quality.

2. Material and Methods

2.1. Input Data

For the analysis of agricultural land use, the IACS (Integrated Administration and Control System) is a very detailed data source. It is available for project purposes for several years for seven Federal Laender of Germany (see Table 1). Since 2005, the IACS combines GIS data on the field scale with land use and land use related information. Because of its application for the administration of agricultural subsidies, the IACS contains only areas for which subsidies have been claimed. Outside this setting, information about land use is complemented by the German Digital Basic Landscape Model (Basis-DLM) which provides nationwide high resolution data with an up-to-dateness of 5 years (BKG 2011). The years 2000, 2005 and 2010 of the Basis-DLM have been selected as input datasets.

As measures of the natural conditions influencing the agricultural land use, general soil maps on different scales, a digital elevation model (DEM, 25 m raster data) and 1 km raster climate data (precipitation, temperature, sunshine duration, transpiration) are available. Within certain protected areas, e.g. nature protection areas, drinking water protection areas and flood areas, land use is legally restricted. In order to identify areas of special sensibility and high nature value, we obtained datasets on organic soils as well as habitat and biotope types.

Most input datasets are vector data and are available as shapefiles. The climate raster has ASCII format and the DEM raster has GRID format.

The available official datasets derive from many different sources due to the federal system and the distinct topics related to specialized governmental organizations. Hence, datasets differ with regard to accuracy, data format and significance of the contained attribute information. As a result, every dataset has to be treated individually.

The growing number of biogas plants represents an economic incentive for the cultivation of biogas maize (Osterburg 2010). According to the Renewable Energy Law (EEG 2012), the Federal Network Agency annually provides a list of all biogas plants, including information on capacity, produced energy and location. For the described case, we applied the 2009 list. Furthermore, we included a dataset with georeferenced nationwide address data in order to integrate the address data of the biogas plants into the GIS database.

Table 1: Input Geodatasets.

Dataset, year of publication, availability	**Data source**	**Data volume (number of geometries)**
Digital Basic Landscape Model (avail. for all Laender) 2000 2005 2010	BKG	up to ~2 Mio. per year and Federal Land
IACS 2003 (RP) 2005 (BB, BY, NI, NW, RP) 2006 (BB, BY, SH) 2007 (BB, BY, NI, NW, RP) 2008 (BB, BY, MV, NI) 2009 (NI, SH) 2010 (SH)	 State Ministries of Agriculture	 up to ~2 Mio. per year and Federal Land
Soil maps Soil map of Germany 1:1 100 000 (avail. for all Laender) Soil map 1:300 000 (BB) Soil map 1:200 000 (incomplete available for all Laender, complete for SH) General geological map of Germany 1:200 000 (available for all Laender) Soil map 1:50 000 (NW, NI) Medium-scale agricultural mapping 1:100 000 (BB, MV) Maps of geomorphology, soil function and soil potential (MV) Soil inventory map (SH)	 BGR LBGR BGR BGR NW: NRW Geological Service, NI: LBEG BB: LBGR, MV: LUNG LUNG LLUR	 5 681 (nationwide) 11 534 11 352 (SH) 270 288 (nationwide) 138 247 (NW) 90 316 (NI) 22 076 (BB) 19 338 (MV) up to ~12 000 364 367
Digital elevation model (available for all Laender)	BKG	~888 Mio. raster tiles
Raster climate data (available for all Laender)	German Weather Service (DWD)	566 364
Protected areas *Water protection areas* Drinking water protection areas Mineral spring protection areas *Flood areas* *Nature protection areas* Nationwide dataset of several conservation areas (sites of Flora-Fauna-Habitat Directive, special protection areas for birds, nature protection areas, landscape conservation areas, natural reserves, national parks, biosphere reserves) Biotope and biotope type mapping, habitat types of Flora-Fauna-Habitat Directive (BB, BY, NI, NW, RP, SH) Peatland maps (BB, BY, MV) Areas related to European Water Framework Directive (BB, NI)	 Various federal State Offices Various federal State Offices Federal Agency for Nature Conservation (BfN) Various federal State Offices Various federal State Offices Various federal State Offices	 up to ~10 000 per Federal Land up to ~48 000 per Federal Land ~1 Mio. up to ~1 Mio. per Federal Land up to 33 550 5341
Administrative units (2008)	BKG	12 533
List of biogas plants (nationwide, 2010)	Federal Network Agency	920 878 lines
Georeferenced Address Database (2010)	BKG	22 Mio.

Abbreviations:

BB *Brandenburg*
BY *Bavaria*
MV *Mecklenburg-West Pomerania*
NI *Lower Saxony*
NW *North Rhine Westphalia*
RP *Rhineland-Palatinate*
SH *Schleswig-Holstein*
BKG *Federal Agency for Cartography and Geodesy*
BGR *Federal Institute for Geosciences and Natural Resources*
LBGR *State Office for Mining, Geology and Minerals of Brandenburg*
LBEG *State Authority for Mining, Energy and Geology Lower Saxony*
LUNG *State Agency for the Environment, Nature Conservation & Geology Mecklenburg-West Pomerania*
LLUR *State Agency for Agriculture, Environment and Rural Areas Schleswig-Holstein*

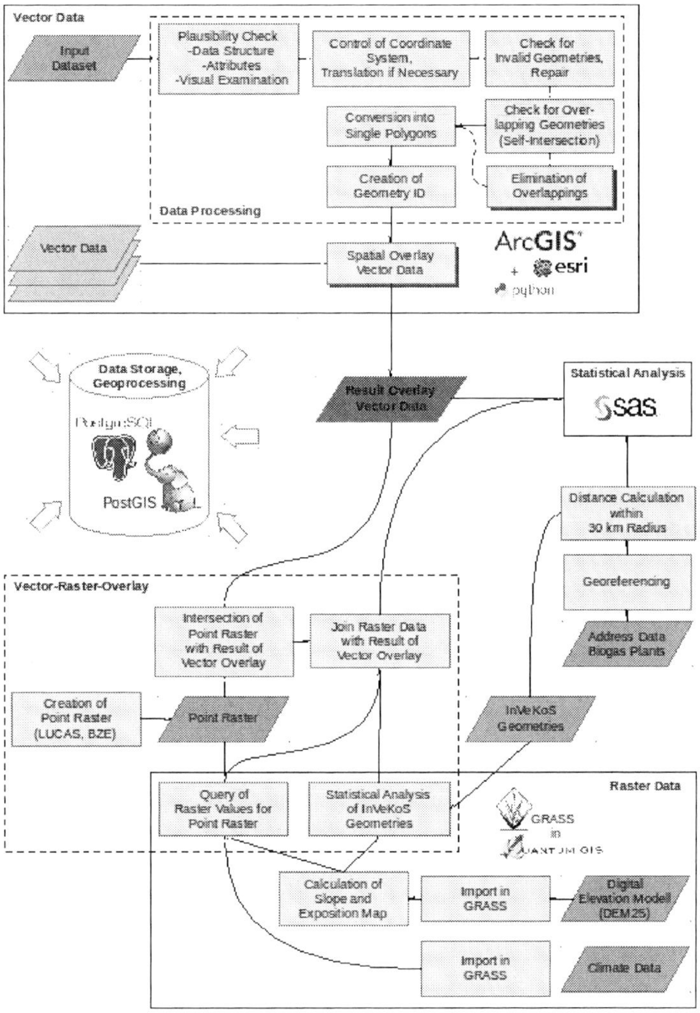

Figure 1: Schematic Overview on the Processing Workflow and the Used Software Products.

2.2. Data Processing – Conceptual Approach

The processing of GIS data is intended to integrate all available and useful information into a combined database ready for statistical analysis. Because of the differing data models of vector and raster data, processing divides into a vector part and a raster part at first and requires a combination of both for full integration.

Most input data are vector data. This and the aim to preserve as much accuracy as possible lead to the decision to use a spatial overlay of all vector data as the main database. Area sums calculated for the resulting geometries build the basis for statistical analysis. Due to this decision, the input datasets have to meet a set of basic conditions. At first, all datasets must not contain larger overlappings, because they would generate errors in area calculation. Further, the extent and content of any dataset has to be complete within nation boundaries or those of the analyzed Federal Laender. With respect to processability, it has to be assured that the vector geometries are valid.

Besides the cleaning of overlapping polygons and the repairing of invalid geometries, standard data preparation procedures include transformation of multi polygons into single polygons for identification purposes and harmonization of spatial reference systems. Figure 1 presents an overview scheme of all processing steps.

Another issue deriving mainly from the heterogeneity of data relates to the appearance of so called sliver polygons – small, elongated polygons at the borders of overlaying polygons with similar but not identical boundaries (see Section 3.2.3).

After the processing of vector data is finished, we have to link the information of the raster datasets to the resulting database (see Section 3.4.2). Additional information such as the situation of biogas plants are integrated into the database by their relation to certain datasets like the IACS data (see Section 3.4.1).

The results chapter describes the implementation of the most important of these steps in more detail with a focus on the implementation in an Open Source environment.

As a part of the statistical analysis, a relation is established between areal values resulting from GIS processing and useful attribute data. In order to identify transition areas where agricultural land use changes have taken place, an explicit assignment and grouping of spatially related IACS land parcels is necessary for the comparison of different years.

2.3. Used Software Products

Not only the number of datasets, but also the size of some of the input data and the resulting database make processing challenging. During consecutive projects, we tested different approaches. Proprietary ESRI product ArcGIS being the standard software in use at our research facility, it was our starting point and our reference for testing. Demands for alternative software resulted from insufficiencies in table management (laborious and very slow at massive data), from increasing failures at overlay operations for massive geodata without the possibility to track the causes and to find a workaround and from slow operation times concerning the combined processing of raster and vector data. The already high license costs of ArcGIS lead to a focus on Open Source software at the search for alternative solutions.

Currently, the data is stored in a PostgreSQL database with a PostGIS extension (PostgreSQL 8.4.5, PostGIS 1.5). This solution guaranties high-performance data access to all project team members while enabling them to execute geoprocessing within the database.

GRASS 6.4, mostly accessed by us in its form as a QuantumGIS extension, is a Open Source software specialized in high-performance raster processing. Therefore, we chose this environment to handle raster data sets like the digital elevation model. In different data manipulation steps it is possible to link the existing raster data with the available vector data.

Even if the mentioned Open Source alternatives work really well, we prefer proprietary ESRI product ArcGIS for combining all vector data available for a Federal Land with a spatial overlay. Until now, this is the most effective possibility to handle this manipulation step (see Section 3.3). During project time, an upgrade from ArcGIS 9 to version 10.1 has taken place. For statistical analysis, we use proprietary SAS 9.2.

3. Results

3.1. Preliminary Results regarding the GIS Toolbox Concept

Concerning the general development and architecture of the the GIS Toolbox, we pursue a strategy of handling the heterogeneous massive geodata within an adequate time frame. The individual approach depends on the requirements and restrictions defined by the project objectives as well as the available data and software.

During development, we often tested different approaches for individual processing modules in order to select the most effective solution (see following sections). In order to limit development effort, we preferred solutions existent or realizable within our software infrastructure (ArcGIS, PostgreSQL/-PostGIS). Criteria for the evaluation of the approaches include feasibility, satisfactory results, processing time and adaptability (for additional datasets or updates). At these decisions, the interactions between software features and the different possibilities for the implementation of certain processing steps have to be considered. Based on the subsequent projects, we advanced the concept of the GIS toolbox continuously.

The system design visualized in Figure 1 therefore represents a preliminary result of the current state and will possibly be enhanced in future projects.

3.2. Elements of GIS Toolbox – Data Preparation

3.2.1. Repair of Invalid Geometries

Invalid geometries occur in many original input datasets because of digitalization errors and different definitions of valid geometries in ArcGIS and PostGIS. Further, invalid geometries can result from geoprocessing routines. However, most geoprocessing routines are very sensible to invalid geometries. This necessitates the application of a repair tool, which was not available as a genuine function until PostGIS 2.0.

In general, approaches like *ST_Buffer*, *ST_SnapToGrid* or the *CleanGeometry* pl/pgsql function created by Horst Düster (2008) were suggested for repairing invalid polygons. Although these approaches work for a variety of invalid polygons, they fail in some cases producing empty or NULL geometries. Therefore, we extended the *CleanGeometry* pl/pgsql function (Düster 2008). The new function combines the described approaches and complements them with additional repair operations like reordering and eventually selecting the nodes for polygon build. Further, we implemented success testing in order to avoid empty geometry entries. Unfortunately, there are still cases, which the function is unable to handle properly. The most frequent invalidity reason is wrong ring order (interior ring is listed before the exterior ring). A plausibility check of the results is therefore necessary to ensure the reliability of the dataset.

3.2.2. Overlapping Polygons

The elimination of overlapping polygons is an important and time-consuming preparation step. We start the procedure with a self-intersection of the dataset, resulting in the identification of all overlappings within it. Only overlappings larger than 1 m^2 are selected for further processing. They are analyzed regarding the reasons for overlapping and the possibilities for their revision. In general, the following types of overlapping polygons occur:

1. duplicates:
    (a) complete duplicates – geometry and attributes, both are similar
    (b) geometrical duplicates – geometries correspond, but attributes are different
2. partly overlapping polygons
    (a) corresponding attributes
    (b) differing attributes

In case of complete duplicates (a.1), we delete one of the polygons. Partly overlapping polygons with corresponding attributes (b.1) are aggregated with PostGIS *ST_Union*. If the attributes of the overlapping polygons differ (a.2, b.2), another approach is used: First, we analyze the attribute data to figure out if some sort of priority ranking exists enabling us to prefer one geometry. This could concern the zoning in a drinking water protection area, or timestamps in the attribute table making it possible to identify the current or valid entry. Then, with *DELETE* (geometrical duplicates (a.2)) or *ST_Difference* (partial overlappings (b.2)), we cut off geometries of lower priority. However, frequently no information on priorities is provided helping to select which geometries to keep. Then, for partial overlappings (b.2), the only solution is to aggregate them with *ST_Union*. Geometrical duplicates (a.2) are reduced to a single geometry without attributes. We document the relation between the resulting geometries and the original attributes in a translation table. This attribute related approach does implicate the design of tailor-made solutions for almost every dataset.

3.2.3. Removing Sliver Polygons in a Timeline Dataset

As mentioned, the IACS datasets are available for several years making it possible to create a timeline of agricultural land use. According to changes in ownership or usage structure and as a result of the correction of the digitalized field borders, the polygons are not static over time. The shifting of field borders between the years leads to numerous sliver polygons after spatial overlay.

Sliver polygons expand a dataset without really adding information. Furthermore, they are very vulnerable for geoprocessing exceptions in later overlay operations. Therefore, we developed a pl/pgsql-function with the aim to identify and eliminate sliver polygons resulting of a spatial overlay. We tested the function for the Federal Land Lower Saxony for which 4 years of IACS data are available (2005, 2007, 2008, 2009).

The identification of sliver polygons within the function is based on the compactness index (cmp) describing the relation of the polygon area (A) to the area of a circle with the same perimeter (U) as the polygon (Nakos 2001). It relies on the fact that circles represent the geometric shape with maximized ratio of area to perimeter.

$$cmp = \frac{4\pi A}{U^2} \quad (1)$$

(Nakos 2001)

In the pl/pgsql function, polygons are characterized as sliver polygons if their cmp is smaller than 0.13, a value which is derived from the definition of slim, very elongated polygons (Nakos 2001). As additional conditions, the function contains upper and lower boundaries regarding polygon area. Polygons smaller than 10 m^2 are always classified as a sliver polygon. We chose the upper boundary of 150 m^2 to prevent sliver polygon classification for polygons of slim shape but large area.

The elimination of sliver polygons in the IACS timeline relies on the assumption that sliver polygons are mostly the result of corrections of the field border digitalization. In this case, the most recent geometry should form the best representation of the actual field border. Therefore, we adjust sliver polygons according to the course of the border line in the most recent dataset.

In a timeline consisting of four years, the function for the identification and elimination of sliver polygons must be executed for each overlay operation. Although, removing sliver polygons has advantages for further overlay operations and statistical analysis, the pl/pgsql function did not yet become part of the general data processing because of its time consuming application.

3.3. Core of GIS Toolbox - Spatial Overlay of Vector Data

After data preparation, we combine all vector data available for a Federal Land with a spatial overlay. For this operation, we use ArcGIS function *UNION* which implies the export of all datasets in .shp format and reimport of the result into the PostgreSQL database again. As ArcGIS 10.1 provides a PostgreSQL database client library enabling direct communication between ArcGIS geodatabases and PostgreSQL, it has to be tested if this time consuming im- and export proceedings are still necessary in the future.

In the meantime, we conducted experiments to implement the functionality of ArcGIS *UNION* in the PostgreSQL environment using pl/pgsql-functions. The result can be found as one of the examples of spatial SQL in the OSGEO User Wiki (Laggner 2011). It consists of a combination of PostGIS functions together with routines for exception handling and data sub-selections in order to speed up processing time. The approach is based on a discussion in the PostGIS users mailing list (postgis-users 2007) and has proven to be most effective in the current software environment (PostgreSQL 8.4, PostGIS 1.5). The overlay dataset is composed of the results of five separate steps:

- intersection (clips the intersecting parts of both input datasets)
- difference A-B (clips all non-intersecting parts of dataset A belonging to intersecting geometries)
- difference B-A (same as above, but vice versa)
- remaining A (selects all geometries of dataset A which are not intersecting with geometries of dataset B at all)
- remaining B (same as above, for dataset B)

Although the PostGIS/PostgreSQL implementation does work, there are still some issues leading to the preference of ArcGIS *UNION*: First of all, only two datasets can be processed at a time which is wearisome if there are at least 10 datasets per Federal Land. This is aggravated by the effort of repairing the invalid polygons after each overlay step and the special treatment of geometries causing geoprocessing exceptions. Also, the resulting geometries contain more sliver polygons and complex polygon boundaries (e.g., unnecessary linestrings) than ArcGIS *UNION* results because of the very high accuracy of the geoprocessing functions in PostGIS.

For the approach using ArcGIS *UNION*, we developed a Python script to control the geoprocessing. The script analyzes the spatial extent of the input data and subsequently calculates the number of parts to which the data should be broken down in order to optimize processing time. Afterwards, the script manages all necessary geoprocessing steps until the spatial overlay operation is finished.

3.4. Special Extensions of GIS Toolbox - Additional Data

3.4.1 Distance calculation biogas plants – fields

Nationwide address data of biogas plants opens the possibility of considering developments of land use changes resulting from the growing number of biogas plants. Besides the postal address, the dataset does not provide any other georeferenced information (coordinates or geometry). In order to georeference the biogas plants, we link their address data to the BKG database with georeferenced nationwide address data.

We use PostGIS for data manipulation. First, we transform the address data of the biogas plant table into a structure conforming with the table containing the georeferenced addresses. Corrections and error checking are a vital part of this step. Because of the heterogeneous enumeration of various address information within one field (person names, street names, house numbers, plot numbers, etc.), it was impossible to extract precise address information such as street name and house number. Therefore, joining biogas address data with the georeferenced address database is mainly based on postcode and place name. For 35 % of the biogas plants, it was possible to use a substring representing the beginning of the street name as an additional join condition. The joining approach leads to the assignment of clusters of points matching the address join condition (1:n relation). For each biogas plant, we aggregate the matching points with *ST_Collect* and identify the centroid of the resulting multi-point geometry.

Availability of matching street names as well as differences regarding the place names between biogas plant and georeferenced address data are handled by a ranked joining procedure taking information quality into account. In the remaining cases without successful matches, we use the centroid of the postcode areas.

For the distance analysis, we calculate the distances between the agricultural areas enclosed in the IACS of 2009 and the biogas plants within a 30 km and a 15 km radius. The agricultural areas are represented by their centroids in order to obtain the average distance. We create the 30 km and the 15 km radius around IACS centroids with *ST_Buffer* and select the biogas plants located inside with *ST_Intersects*.

3.4.2 Processing of Topographical Raster Data in GRASS

The digital elevation model (DEM) is a 25 m raster dataset providing height information for Germany. In order to get more detailed information about the topography as an important local factor, we link the DEM data with the available vector data. We analyze the topography for three of the original seven Federal Laender (Lower Saxony, Bavaria and North Rhine-Westphalia).

Because of the amount of data, processing is very difficult and we tested various approaches. Currently, we add topographical information to the vector data in two separate processing steps. First, we extract DEM information for the agricultural acreage represented by the most recent IACS geometries on a highly detailed level. We chose a more condensed approach for the areas outside the IACS (point grid described below).

We conduct the data manipulation with GRASS 6.4. Besides altitude, slope and aspect are relevant for agricultural use. We derive raster maps for the two last variables from the DEM with the r.slope.aspect module of GRASS. For the linkage between the topographical raster maps and the IACS polygons, we use the GRASS module *v.rast.stats2*. This module calculates univariate statistics like average, standard deviation, minimum, maximum, median or percentiles for all tiles of a raster map intersecting with overlaid vector polygons. The results are attached to the attribute table of the vector map.

In order to complement topographical information for the areas outside the IACS, we chose a different procedure. We generate a Germany-wide regular 100 m vector point grid. The points are based on the 1 km LUCAS grid (Eurostat 2013) and were densified to a 100 m grid. This approach intents to generate a sufficient number of samples for statistical analysis. Furthermore, we add additional points in order to meet the known sampling points of the German agricultural soil inventory (TI-AK 2013). This will allow to easily link information of the soil inventory in future.

We intersect the point grid with the polygons of the vector overlay in order to take a spatial sample of this pattern of land use and local conditions for the three Federal Laender which are topographically analyzed. In the next step, we attach the topographical information of the raster maps to each point of the selected grid matrix. We use GRASS (*r.mapcalc*) to combine vector and raster data.

3.5. Use Case: Biogas-induced Land Use Change Between 2005 and 2007 as an application for GIS approach

A recent development in Germany is the change in land use because of the expansion of biogas production based on renewable resources from agriculture. In this respect, an increase of the cultivation of maize as the dominant fermentation substrate was observed which is attended by a conversion of permanent grassland to agricultural cropland. This causes concerns regarding the impacts on nature and water protection goals which were basic objectives in the WAgriCo2 project (follow-up of EU-LIFE project „Water Resources Management in Cooperation with Agriculture"). The following findings are excerpts of the final project report (Osterburg 2010).

The changes in land use regarding the trends of maize acreage and the conversion of permanent grassland are analyzed for Lower Saxony based on IACS data for the years 2005 and 2007. The IACS data is set in relation to local characteristics such as soil types, topography, the establishment of nature and water protection areas as well as the distance of the land parcels to biogas plants.

For the statistical analysis, land use changes are quantified as net changes because otherwise normal crop rotation would increase the total area evolved without reflecting real change.

Between 2005 and 2007, the area of maize increased by approx. 60 000 ha in Lower Saxony. The IACS data enables further analysis of these changes classified according to farm or land parcel properties.

From the distance analysis, an agglomeration effect for biogas maize cultivation in the vicinity of biogas plants is observable. 74 % of all biogas maize is cultivated within a radius of 5 km from a biogas plant, (95 % within 8 km radius).

Analyses of correlations with moisture classes and potential soil erosion caused by precipitation show expansion trends of biogas maize cultivation into sensitive areas. The expansions into erosive areas as well as into wet areas have to be considered as a potential concern regarding the future water and soil protection. The intensification of pressure on the wet areas is of special relevance because the often organic soils contribute to greenhouse gas emissions if the organic carbon is mineralized under intensive agricultural use (Nitsch et al. 2012).

Within the same period, a net shift of approx. 19 000 ha from permanent grassland to agricultural cropland has taken place in Lower Saxony. Maize has, with 59 %, the highest percentage of the cultures following the ploughing of grassland. Still, biogas maize is responsible for only 8 % of the conversion areas while other maize comprises the remaining 51 %.

When analyzing the protective effect of reserved areas for nature and drinking water protection aims, significant effects can be found on areas where grassland protection regulation is implemented and where participation in agri-environment measures takes place more frequently. It could be assumed, that some protection effects are caused by already less suitable agricultural conditions on the protected areas.

4. Discussion

The selection and development of adequate processing methods for relevant GIS data is defined as an important part of the research interests within all mentioned projects. The first part of the discussion therefore addresses the appropriateness of the chosen approach including the contribution of the Open Source software and examines possible future options. In the second part, the discussion focuses on the findings of the 'use case' amongst other things pointing to some data uncertainties.

4.1. Evaluation of GIS Approach

We can state that the current geoprocessing approach leads to the aimed result of comprehensive GIS datasets ready for statistical analysis. However, experiences to date reveal the enormous effort needed for data preparation and processing. With a growing number of datasets through further updates with additional years, the overlay operation becomes more time consuming and results in a larger dataset. This means, the amount of information which can be analyzed together is limited. Different approaches for future handling are possible:

First, the dataset could be more or less fixed for a certain date with long update cycles. During an update, old contents are replaced by new ones in order to avoid excessive database expansion. Apart from the reduced usability because of possibly out-dated information, with this approach the complete, time consuming overlay operation might become necessary on a regular basis.

Second, all available datasets are prepared but overlay operations only take place for specific analyses with the needed subset of data. Advantages would derive from the highest achievable timeliness of data and the minimization of data size per analysis. Hence, this approach is only reasonable, if small data subsets are sufficient for most research questions. If too many datasets are to be included, the subsequent effort could in total even exceed that of the current approach.

Third, a complete change of the approach could be conceivable from the hitherto spatially inclusive, polygonal approach to a statistical sample based on a regular point grid. This approach would offer many advantages for the geoprocessing. It would dramatically reduce the effort of the overlay operation because no complicated polygons have to be processed. Topology exceptions due to invalid input or output polygons are avoided. Adding to convenience, the elimination of overlappings is limited to those affecting the point grid, and reference tables suffice as processing routines regarding the geometries. Furthermore, integration of raster data is easily manageable. For more detailed research aspects, especially with respect to IACS data, the point grid should be complemented with polygonal data in a way corresponding to the second approach.

Which approach will be pursued later on has to be decided within the context of future research projects also keeping software improvements in mind.

With respect to the Open Source software which was used for geoprocessing, it has been proven that it is suited for the processing of massive geodata. The processing steps executed with Open Source software were always finished within hours or days, which is acceptable seen the bulk of data. As a special strength, the PostgreSQL database facilitates storage and manipulation of many datasets, enabling data access for all project team members. PostGIS, the spatial extension for PostgreSQL, provides many geoprocessing functions. Compared to ArcGIS, the numerous possibilities to develop workarounds whenever the preferred solution fails, are an important advantage of PostGIS.

One major drawback of PostGIS is its sensitivity for topology exceptions. The instances of failing operations increase with growing numbers of geometries to be processed. In some cases, this requires a massive use of error trapping routines expanding the geoprocessing operations considerably. The sensitivity for topology exceptions is associated with the high accurateness of some geoprocessing functions. Thus, the addition of a scalable tolerance enabling the user to consider the resolution of a dataset would represent a desirable improvement.

A combination of different software is necessary at the given stage of used software products. The data transfer between them causes additional effort and parallel data storage. A better data interchange with the PostgreSQL database is established with the current version of ArcGIS 10.1. This enables the user to directly access the PostgreSQL database as a data source and storage location for inputs and results of geoprocessing routines in ArcGIS. In addition, new features in PostGIS 2.0, like the implementation of raster data processing and topological models for vector data, open prospects for possible future method adaptations.

A further point for discussion is the informative value of the input datasets. Having IACS data for seven Federal Laender at our disposal, a comparison of the results and the composition of a nationwide database would be desirable. Due to differences in availability between the Federal Laender, the overlay results are unique considering the integrated IACS years as well as specific input data. The latter is not only related to thematically specific datasets (e.g., peatland mappings) but also to the spatial resolution of datasets issuing the same subject. A good example is the variety of available soil maps (see Table 1). The only soil map with nationwide extent is the Soil Map of Germany 1 : 1 000 000. The Soil Map of Germany 1 : 200 000 is in preparation, but is not available in full coverage. Therefore, it is appropriate only for some of the Federal Laender. For some soil related issues, the Geological Map of Germany 1 : 200 000 is the best data source at national scale. Soil maps with finer resolution are only available from federal institutions and the acquired datasets differ in resolution (e.g., 1 : 300 000, 1 : 100 000, 1 : 50 000) and included attribute information.

Resolution differences arise between most of the input datasets. This is inevitable, if information regarding all relevant topics has to be assembled. In consequence, inaccuracies are to be expected from the result of the overlay operation, accumulating especially near polygon borders.

As another issue, input data represent information for a specific period in time. This means that they are limited concerning their comparability for different Federal Laender and the timeliness of their information.

The inaccuracies deriving from georeferencing have to be considered when working with the dataset of the biogas plants. Besides the described difficulties in matching the address data with the georeferenced dataset, the postal address of a biogas plant may not be consistent with its actual location. The generated point geometry can only be an approximation of the actual address.

4.2. Discussion with Respect to 'Use Case' Results

The analysis of land use changes induced by increased biogas production as presented in Chapter 4 bases substantially on the interpretation of IACS data. The main contribution of analyzing the data within a GIS consists in the spatial link of datasets with different subjects or time-references, allowing the identification of causal connections and temporal changes.

IACS provides the most detailed data available with respect to land use in agricultural areas. It contains information regarding the applying farm, the agricultural use and the applied subsidies. The land parcels are mapped with high accuracy because their extension has an effect on the granted subsidies (EC 2004). Nevertheless, the accuracy of IACS data regarding the identification of land use changes is limited.

Many IACS land parcels are aggregates of more than one field with different crops and owners. This leads to a certain fuzziness on the spatial allocation of land use changes. The identification of changes within a land parcel is only possible if an areal difference is reported regarding the agrarian use for this land parcel between 2005 and 2007 (Osterburg 2010). Land use shifts of a corresponding extent within a land parcel remain unnoticed. The described land use changes therefore represent minimum changes. However, the error induced by this inaccuracy is estimated to be negligible because of the small size of land parcels (3 ha on average) in Lower Saxony (Osterburg 2010).

Similar to the localization of land use changes within a land parcel, it is also difficult to track farm-related management changes on land parcels where more than one farm operates.

Furthermore, the differentiation between biogas maize and other maize can only be based on the information that energy crop aid was applied for a land parcel. In comparison with additional statistical data, 20 % of the expected biogas maize acreage could not be identified in the IACS data (Osterburg 2010).

As already discussed in Section 4.1, the georeferencing of the biogas plants is afflicted with uncertainties which are transmitted to the results of the distance calculation. Since the analysis shows the highest effect in the immediate vicinity of biogas plants, it should be kept in mind that in some cases the georeferencing error might be higher than the calculated distance.

Although the presented 'use case' focuses on a specific research aspect, the already developed database and GIS toolbox provide a very useful basis for further land use related research. Corresponding to the research interests of the Thünen-Institute of Rural Studies, other use cases might include the agricultural use of organic soils and the related effects on greenhouse gas emissions or the problem of nutrient contamination in groundwater and surface water.

5. Summary

The described geoprocessing approach is sufficient for handling massive and heterogeneous geodata and provides the targeted results. Nevertheless, the continuation of the approach is limited because of the long processing time and the massive amount of data resulting from data acquisition and processing. The requirements of future projects will influence the decision on how the approach will change for their purposes.

Open Source software is able to substitute proprietary software like ArcGIS in most processing steps. The main objection concerning the performance of the used PostGIS version relates to the topological sensitivity of its geoprocessing functions.

Statistical analysis based on the processed geodata allows reliable statements regarding relevant phenomena in agricultural land use change. Hence, it has to be taken into account, that the heterogeneity of the available geodata together with the uncertainties inherent to the input data causes inaccuracies in the spatial overlay and the resulting area calculation.

The use case demonstrates how relevant variables can be analyzed together based on the available data and which conclusions can be drawn from the results. There is an increase in biogas maize cultivation probably leading to a regional reorganization. Direct cumulative effects of biogas plants are observed mainly within very short distances. The gain in biogas maize cultivation contributes to the pressure on sensible and valuable areas. A protective effect is strongest if explicit regulations exist regarding agricultural land use.

The use of the database and the GIS toolbox is not limited to a specific research aspect but can be transferred to other land use related issues.

Acknowledgement Some of the findings presented here were produced during EU-LIFE project WagriCo2 („Water Resources Management in Cooperation with Agriculture", part II) carried out at the Thünen-Institute of Rural Studies. Results from statistical analysis are cited from the project work report (Osterburg 2010). Furthermore, we want to mention the project funded by the Federal Agency for Nature Conservation (BfN) "Evaluation of the Common Agricultural Policy from a nature conservation point of view" (2008-2009) and the cooperation project with the Thünen-Institute of Climate-smart-agriculture "Analysis of land use change and development of methods for identification and quantification of measures for greenhouse gas abatement in the agricultural sector (part I and II)", which also contributed to the process of method development.

References

Düster, H. (2008), 'cleanGeometry - remove self- and ring-selfintersections from input Polygon geometries'. URL: http://www.kappasys.ch/pgtools/cleangeometry/cleanGeometry.sql (Download: 01.09.2008, Last Checked: 29.01.2013)

EEG (2012), 'Erneuerbare-Energien-Gesetz vom 25. Oktober 2008 (BGBl. I S. 2074), das durch Artikel 5 des Gesetzes vom 20. Dezember 2012 (BGBl. I S. 2730) geändert worden ist'. URL: http://www.gesetze-im-internet.de/bundesrecht/eeg_2009/gesamt.pdf (Last Checked: 06.02.2013)

European Commission (2004), 'Commission Regulation (EC) No 796/2004 of 21 April 2004 laying down detailed rules for the implementation of cross-compliance, modulation and the integrated administration and control system provided for in of Council Regulation (EC) No 1782/2003 establishing common rules for direct support schemes under the common agricultural policy and establishing certain support schemes for farmers'. URL: http://eur-lex.europa.eu/LexUriServ/LexUriServ.do?uri=CELEX:32004R0796:en:NOT (Last Checked: 29.01.2013)

Eurostat (Last Checked: 29.01.2013), 'Land use/cover area frame survey (LUCAS) - Introduction'. URL: http://epp.eurostat.ec.europa.eu/portal/page/portal/lucas/introduction

BKG (Federal Agency for Cartography and Geodesy) (2011), *Digital Base Landscape Model – Basis-DLM*, Status of documentation: 20.09.2011. URL: http://www.geodatenzentrum.de/docpdf/basis-dlm_eng.pdf (Download: 29.01.2013)

Laggner, B. (2011), 'Set of functions for spatial overlay of 2 polygon tables'. URL: http://trac.osgeo.org/postgis/wiki/UsersWikiExamplesOverlayTables2 (Last Checked: 29.01.2013)

Laggner, B., Orthen, N. (2013), 'Verwendung räumlicher Massendaten bei der Analyse von Landnutzungsänderungen', *in proceedings of* 'Massendaten-management in der Agrar- und Ernährungswirtschaft, 33. GIL – Jahrestagung, 20. - 21. Februar 2013, Potsdam', pp.183–186.

Nakos, B. (2001), 'On the Assessment of Manual Line Simplification Based on Sliver Polygon Shape-Analysis', *in* '4th Workshop on Progress in Automated Map Generalisation (ICA, Beijing, China, August 2001)'.

Nitsch, H., Osterburg, B., Roggendorf, W., Laggner, B. (2012), 'Cross compliance and the protection of grassland – Illustrative analysis of land use transitions between permanent grassland and arable land in German regions', *Land Use Policy* 29, 440-448.

postgis-users (2007), 'Union of 7 datasets', thread in postgis-users mailing list archive. URL: http://lists.osgeo.org/pipermail/postgis-users/2007-August/016804.html (Last Checked: 06.02.2013)

Osterburg, B. (2010), 'Wirkungen von Biogasanlagen auf die Flächennutzung in Niedersachsen – eine Analyse von Daten des Integrierten Verwaltungs- und Kontrollsystems (InVeKoS)', Arbeitsberichte aus der vTI-Agrarökonomie 05/2010, Braunschweig, September 2010.

Schramek, J., Osterburg, B., Kasperczyk, N., Nitsch, H., Wolff, A., Weis, M., Hülemeyer, K. (2012), 'Vorschläge zur Ausgestaltung von Instrumenten für einen effektiven Schutz von Dauergrünland', *BfN-Scripten* 323, Federal Agency for Nature Conservation (BfN), Bonn.

TI-AK (Thünen-Institute of Climate-smart-agriculture), 'German agricultural soil inventory'. URL: http://www.ti.bund.de/en/startseite/institutes/climate-smart-agriculture/research-projects/german-agricultural-soil-inventory.html (Last Checked: 29.01.2013)

Open Source in accessibility analyses

Modeling of street petrol station accessibility in Germany

by Stefan Neumeier

Johann Heinrich von Thünen-Institut, Federal Research Institute for Rural Areas, Forestry and Fisheries, Germany. stefan.neumeier@ti.bund.de

Abstract

The paper relates two closely intertwined story lines. A socioeconomic one concentrating on street petrol station accessibility in Germany's rural areas and a geospatial one concentrating on the usability of the Open Street Map within rural studies. This is attributed to the fact that the paper builds on findings from applied research within rural studies and is intended to serve as a "practical experiences report" assessing the usability of open source GIS/-data in rural studies. Here we analysed the accessibility of street petrol stations as one core service of general interest important for the overall individual mobility of the population, especially in rural areas, based on an exemplary raster-based GIS accessibility analysis. This analysis builds upon an open source approach using PostgreSQL/PostGIS as well as the Dijkstra shortest path algorithm implemented in the Perl-module „Graph-0.94". Besides acquiring objective data on street petrol stations accessibility for policy advice, we were also interested in reviewing the usability of OpenStreetMap (OSM) data compared to commercial routing networks (ESM).

Altogether the findings suggest that in Germany, street petrol stations are for the majority of the population (ESM: 99.5%/ OSM: 99.4%) quite accessible. On average the distance to the next street petrol station amounts 5.4 km (ESM)/ 5.5 km (OSM). Regions with disadvantageous accessibility are predominantly sparsely populated.

The comparison of the accessibility values calculated based on ESM and OSM showed that great differences exist in a per cell comparison as well as on the community level, whereas aggregated average accessibility values for greater aggregates like counties proved to be comparable to each other. Against the background of OSM's lower level of completeness in rural areas an interesting but unexpected result is the fact that the accessibility values differ within urban as well as rural areas within the same range. Nevertheless, considering the identified shortcomings of the OSM, the data set still seems to show a lower performance than commercial data sets.

Keywords: Petrol Station Accessibility, Shortest Path Analysis, OpenStreetMap.

1. Introduction

Amongst other factors, accessibility is one precondition for the economic, social and cultural development of regions (Hemetsberger & Ortner 2008). For example, good accessibility is one precondition for the participation of a region's resident in the economic prosperity of the centres; for participation in potential development opportunities, and the economic capacity of business locations (Hemetsberger & Ortner 2008, Platzer & Gmeinhart 2003). However, accessibility is not only important for location decisions and regional development but also for the individual life situation of the citizens. The quality of infrastructure accessibility also determines the regional infrastructure supply (BMVBS: Raumordnungsbericht 2011, p. 55). For the current discussion on ensuring the provision of services of general interest (SGI) for Germany's rural areas, held against the background of the demographic change and its potential consequences as well as the normative aim of providing comparable living conditions in all areas, up-to-date information about the accessibility of SGI is important in order to get an objective and realistic view on the current situation that can function as input for policy interventions (Schulz & Bröcker 2007). Accessibility information can also be utilized as input for catchment analyses, potential analyses and econometric analyses, for the building of regional typologies or for regional benchmarking. Objective methods of assessment and comparison are needed for this purpose. Data aggregated in statistical regions (supply indicators) strongly distort the situation as important intraregional disparities are levelled by the aggregation. Activity-based accessibility indicators like the accessibility of regional metropolitan areas mainly permit only indirect conclusions due to the accessibility of SGI (e.g., based on the level of centrality of the different places and their planned provision of infrastructure). Therefore more concrete data on the provision of SGI, is needed to enable scientists as well as politicians to get an idea of the true situation. The modelling of accessibility situations based on a raster-based accessibility analysis is one solution for obtaining realistic quantitative accessibility data. This information will be presented in the next paragraphs, taking as example the accessibility of street petrol stationsin Germany. The analysis itself is based on an open source approach using PostgreSQL/PostGIS as well as the Perl-module „Graph-0.94" and is based on the Dijkstra shortest path algorithm In addition to acquiring objective accessibility data for policy advice, the study also aims to compare the usability of the OpenStreetMap (OSM) data as a low/no cost alternative for rural studies, that according to Ludwig et al. (2010) is less complete in comparison to commercial routing networks especially for rural areas. Thus, instead of concentrating on pure GIS technical facts only, the article purposely relates two story lines a socioeconomic one (street petrol station accessibility in Germany's rural areas) and a geospatial one (usability/constraints of the Open Street Map within rural studies), that are closely intertwined. This is attributed to the fact, that the article builds on finding from applied research within rural studies and is intended to serve as a "practical experiences report".

The remainder of the article is divided into seven chapters. In sections 2 to 4, key data are introduced on the situation and importance of street petrol station accessibility as well as the core concepts of accessibility. In sections 5 the methodology of the accessibility analysis is explained. In sections 6 key results are presented comparing accessibility data acquired by using commercial street data and data acquired by using OSM data. In section 7 the lessons learned due to the street networks as well as software used are summarized. The paper concludes with a consideration of the usability of OSM data in research focusing on rural areas.

2. Key data to mobility in rural areas

Mobility is generally defined as movement, whereas a differentiation is made between social mobility (change in the socio-economic status) and spatial, respectively geographical, mobility (migration, traffic, transport) (Leser et al. 1993, p. 409, Johnston et al. 2000, p. 507). The article focuses on geographic mobility. The development of the mobility behaviour is characterised by a rising motorisation of private households; an increasing share of non-commercial traffic; a more and more circadian traffic situation; higher growth rates in shopping and leisure time traffic, as well as increasing time budgets for daily mobility resulting in increasing travel distances (Küpper & Steinrück 2010, p. 14). Reasons are the individualisation, orientation towards leisure-time of society, increasing household incomes, increasing distances between place of residence and the places of other functions of daily life as well as traffic policy and policy areas influencing spatial and settlement patterns (Küpper & Steinrück 2010, p. 14).

Generally only rudimentary public transportation exists in rural areas or sub-optimal station times prevail so that citizens are dependent on their cars if they want to be mobile. Additionally, especially in rural areas, comparatively long distances have to be covered to reach, for example, the next supermarket, place of employment or medical services (Küpper & Steinrück 2010, p. 5). In sparsely populated rural areas nearly two thirds of all travel is covered with passenger cars (Küpper & Steinrück 2010, p. 14).

Therefore, according to Heinze et al. (1982), the availability of a passenger car is a key to individual space and time structuring (Heinze et al. 1982, p. 368). But in order to be mobile, not only the availability of a passenger car, but also the accessibility of a petrol station, is important.

3. Key data to the German petrol station market

Petrol stations can be attributed to the local supply, meaning goods and services for short- and mid-term needs near to the place of residence. They can be seen as services complementing the basic services. In Germany there were 14,744 petrol stations in 2011. These petrol stations can be divided into 377 so called highway petrol stations[19] and 14,367 street petrol stations (ADAC 2012).

According to a study about the petrol station market, Germany, - with less than two petrol stations per 10,000 citizens and three petrol stations per 10,000 cars -. belongs to the countries with the lowest petrol station densities in Europe (Morgenstern & Zimmermann 2012, p. 15). Besides selling petrol, a large number of petrol stations in Germany complement their core business with additional retail trade (Korn 2006, Morgenstern & Zimmermann 2012).

In order to keep services competitive in rural areas they are often concentrated in central locations at the expense of peripheral regions (OECD 2007, p.

[19]This are petrol stations that are only accessibly by highway, including the outlets of the mineral oil companies.

18-19). This also applies to the provision of petrol near the place of residence. Since 1970 the number of street petrol stations in Germany has decreased from 46,684 street petrol stations in 1970, to 14,367 street petrol stations in 2011 (ADAC 2012). The thinning of the petrol station network led to an increasing spatial concentration at highly frequented profitable locations (Gyllensvärd 1999, Korn, 2006). Unfortunately regionalised data on street petrol stations is not available, but considering the trend to a thinning of SGI, it can be assumed that especially unprofitable locations with comparatively low numbers of customers, especially in rural areas, have closed.

4. Accessibility

The concept "accessibility" is used in quite a few areas, including infrastructure- and city planning or marketing. Therefore the term "accessibility" has quite a few meanings. But generally accessibility can be understood as the number of opportunities for economic and social life that can be reached with a reasonable effort. Accessibility describes the quality of a point in space constituted by its traffic connections to other attractive points in space. So accessibility is the main product of transport systems (Bleich & Koellreuter 2003, p. 7, Schürmann et al. 1997, Schwarte 2005). Different methods exist to get accessibility values that can roughly be grouped into three categories: (1) approaches common mainly in a regional economy based on spatial interaction models (e.g., gravitation models, logit-models) (Bleisch 2005, Schulz & Bröcker 2007); (2) approaches common in transport sciences based on a prognosis of the traffic situation (e.g., gravitation models, opportunity models, random utility theory) (Bleisch 2005, Schulz & Bröcker 2007) and (3) approaches focusing on the geographic accessibility (Euclidean distance, distance within route networks) (Hemetsberger & Ortner 2008) as the approach followed in the study. One peculiarity of geographic accessibility is that in regions with a dense road network like metropolitan agglomerations and cities, the Euclidean distance provides accurate enough results whereas in regions with sparse road networks like rural areas, the determination of the Euclidean distance is insufficient. This is because the circumvention of natural and anthropogenic barriers is more likely to extend the distances to be covered (Dahlgren 2008, p. 16). In such cases it is therefore better to determine the geographic accessibility based on distances within the traffic network. This kind of analysis can be supported by the use of network analysis methods that are part of many Geographic Information Systems (GIS). Thereby two main approaches can be used:

1. In a more traditional, less cost intensive approach for a selected, quite limited number of starting points (e.g., centroids of polygons representing administrative units) the distance to targets is computed based on a shortest path network analysis – often with predetermined time- cost budgets. The results are catchment areas, respectively time/cost matrices. The main drawbacks of this approach are that the resulting isochores do not allow a further internal differentiation and that results are not spatially inclusive and comprehensive.

2. In a quite cost-intensive raster-based approach the region under consideration is overlain by a raster with a pre-defined feed size. The centroids of the raster cells represent the starting points of the analysis. That means for every raster-centroid, the shortest distance to the next target is computed and the resulting distance value is attributed to the raster cell. The advantage of this approach is that the distance can be computed from almost unlimited starting points (raster cells) to a high number of targets. Results are also time/cost matrices but ones which, in contrast to the former approach, are spatially inclusive and comprehensive. Furthermore the resulting isochores can be further differentiated internally. Based on the single results, calculation of areas can easily be achieved, e.g., by computing the arithmetic mean. The feed size of the raster cell determines the accuracy of the model as the smaller the feed size the more accurate is the model. Simultaneously, the smaller the feed size the more computation costs are necessary. Thus the feed size of the raster cells has to be selected carefully considering the available resources and desired/necessary accuracy of the results.

5. Method: GIS accessibility model

As the raster-based approach is more flexible, it was decided to take this approach as a basis for the calculation of the street petrol station accessibility. All data have been transformed to the geographical reference system „DHDN/3-degree Gauss-Krüger Zone 3" (SRID/EPSG: 31467). The data preparation and analysis was performed with PostgreSQL 9.1.1, PostGIS 1.5.3 and Perl 5.14.2.

Road network

Besides obtaining accessibility information for policy advice, the study also aimed at assessing the usability of OSM data released under the Open Data Commons Open Database Licence (OdbL) as low/no cost alternative compared to commercial street network data taking as example the Esri StreetMap (ESP) dataset. The OSM is a „collaborative project in which volunteers are creating a free and editable map of the world, to reduce the dependency on commercial data providers" (Ludwig et al. 2012, p. 148). That means the data of the OSM is based on data collected by volunteers interested in participating at the project, whereas practically everybody is able to participate. One major drawback inherent to the OSM project is that data collection does not follow a specific acquisition plan, meaning it depends on the volunteers participating for which region data is collected. All in all, Ludwig et al. (2010) showed that in Germany the OSM is less complete compared to commercial data sets, whereas the completeness of the OSM is higher in densely populated areas than in sparsely populated ones (including rural areas) (Ludwig et al. 2010, p. 149). Nevertheless because the OSM can be used without having to pay licence fees, the question arises if despite the drawbacks concerning the overall completeness, especially in rural areas, the dataset can be used for accessibility analyses without distorting the results too much. Therefore, in order to evaluate the usability of OSM for accessibility analyses, the analysis of the street petrol station accessibility has been carried out based on ESP data that are based on data from NAVTEQ and Tele Atlas as well as on OSM data. Thereby the download as well as preparation of the OSM street data was performed with the help of the tool „OSM2PO" developed by Carsten Möller (`http://osm2po.de/`).

Street Petrol Stations

The locations of street petrol stations have been derived from the database „D Tankstellen" that was provided by the POICON GmbH, a company specialized in points of interest for car navigation systems. The street petrol station dataset is provided for free at the POICON website but is copyrighted by the POICON GmbH. It was deliberately decided not to use the petrol station data set of the OSM (amenity fuel) as this data set proved not to meet the requirements desired for following reasons: (1) For Germany the OSM contains only information about 12,971 petrol stations (ca. 87 % of the overall German petrol stations). As no information is given on missing petrol stations it is more than likely that systematic missing locations might occur especially in rural areas which would distort the result of the accessibility analysis considerably. (2) The OSM dataset does not differentiate between the street petrol stations in which we were interested and highway petrol stations, which we wanted to exlude from our study. (3) The OSM data set does not allow explicit differentiation between branded and independent street petrol stations. (4) Experiences showed that amenity tags within the OSM seem not to be quality controlled or validated, rising questions about validity. The POICON data set "D Tankstellen" used contains geographical locations for 14,217[20] street petrol stations in Germany. Altogether the data set contains 149 (ca. 1 %) street petrol stations less than those listed in the current German Petrol Station Statistics 2012. So the data set covers approximately 98.96 % of Germany's street petrol stations.

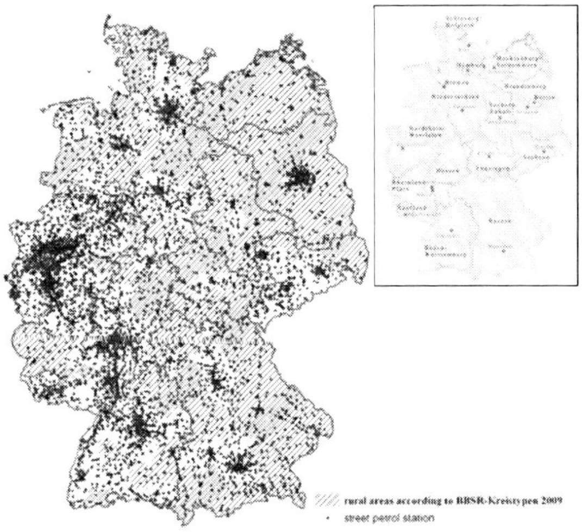

Figure 1: Distribution of street petrol stations in Germany. (Data sources: POICON GmbH; Administrative boundaries: Bundesamt für Kartographie und Geodäsie 2010).

Figure 1 shows the distribution of street petrol stations in Germany. It can clearly be seen that the density of the street petrol station network decreases

[20] Altogether the dataset „D Tankstellen" contains 14,222 street petrol station locations. However, five locations were excluded from the analysis as they proved to be outside Germany.

from the agglomerations to the more rural areas. Furthermore it can also be seen that the density of the network is comparably lower in the new federal states „Mecklenburg-Vorpommern", „Brandenburg", and „Sachsen-Anhalt".

In 2009, the Federal Institute for Research on Building, Urban Affairs and Spatial Development (BBSR) developed the so-called BBSR-Kreistypen 2009 (BBSR 2009a) that are meant to assist in an intraregional comparison of different types of regions within Germany. The system differentiates between cities, defined as urban districts with more than 100,000 citizens and other districts which, including the hinterland, have less than 100,000 citizens. Outside the cities the types are defined based on the population density as well as the settlement structure.

A merging of the street petrol stations with the BBSR-Kreistypen 2009 provides us with a supply indicator due to street petrol stations. This gives us a first impression about potential differences in the street petrol station availability within the different types of regions in Germany, as well as about areas which are potentially disadvantaged in terms of street petrol station accessibility (see Table 1). The resulting data suggests that in Germany, street petrol stations are least accessible in areas identified as rural.

Table 1: Distribution of street petrol stations in Germany according to federal state and BBSR-Kreistyp 2009 (Data sources: POICON GmbH, BBSR).

Federal State	Number of street petrol		street petrol stations per BBSR-Kreistyp 2009																	
			1		2		3		4		5		6		7		8		9	
	total	%	total	%	total	%	total	%	total	%	total	%	total	%	total	%	total	%	total	%
Baden-Württemberg	1650	11.6	191	12	539	33			22	1.3	97	5.9	625	38	176	11				
Bayern	2319	16.3	304	13	91	3.9	220	9.5	42	1.8	141	6.1	357	15	291	13	682	29	191	8.2
Berlin	342	2.4	342	100																
Brandenburg	426	3.0	22	5.2					264	62	15	3.5			67	16			58	14
Bremen	118	0.8	96	81							22	19								
Hamburg	275	1.9	275	100																
Hessen	1136	8.0	181	16	289	25	199	18			31	2.7	232	20	123	11	81	7.1		
Mecklenburg-Vorpommern	294	2.1									22	7.5			35	12	55	19	182	62
Niedersachsen	1472	10.4			214	15	156	11	118	8	159	11	410	28	214	15	138	9.4	63	4.3
Nordrhein-Westfalen	3415	24.0	1220	36	1233	36	291	8.5			63	1.8	507	15	101	3				
Rheinland-Pfalz	703	4.9	29	4.1	38	5.4	81	12			71	10	365	52	119	17				
Saarland	165	1.2	61	37	84	51	20	12												
Sachsen	607	4.3	130	21			112	18	27	4.4	49	8.1	213	35			76	13		
Sachsen-Anhalt	359	2.5									74	21	25	7	168	47	46	13	46	13
Schleswig-Holstein	596	4.2			52	8.7	105	18	39	6.5	86	14	90	15	69	12	83	14	72	12
Thüringen	340	2.4									57	17	76	22	89	26	92	27	26	7.6
Germany	14217	100	2851	20	2540	18	1184	8.3	512	3.6	887	6.2	2900	20	1452	10	1253	8.8	638	4.5

1 Core cities in agglomerations
2 Densely populated distcicts in agglomerations
3 Highly populated districts in agglomerations
4 Rural districts in agglomerations
5 Core cities in urbanised areas
6 Densely populated areas in urbanised areas
7 Rural districts in urbanised areas
8 Densely populated rural areas
9 Sparsely populated rural areas

Raster of Reference

The Austrian Institute of Technology (AIT) - Population Density GRID (Steinocher, Köstl, Weichselbaum, 2001) for Germany has been chosen as base raster for the analysis. The AIT-GRID is based on Eurostat's European geostatistical 1km grid „Grid_ETRS89_LAEA_1k", which was built based on Corine Land Cover 2006 data and which covers areas classified as settlement. The AIT-GRID enhances the European geostatistical 1km grid by adding attribute information due to absolute population according to Eurostat 2006 data for every grid cell. Although the AIT-GRID is based on 2006 population as well as settlement data and does therefore not represent the current state of settlement and population development, it was decided to use this raster as base raster of the accessibility analysis, as it was the only approved data set available with gridded population data for Germany. In order to be inherently consistent it was decided to use the AIT-GRID as is and not to update or merge it with other settlement or population data sets. Altogether the raster contains 209,203 cells[21] with a feed size of 1,000 m. The raster was chosen for two reasons. First, as the main aim of the analysis is to analyse the proximity of street petrol stations to the place of residence or employment, the raster allows computation costs to be reduced by concentrating on the areas of interest -the settlement areas- as starting point only. Second, and more importantly, the grided population data offers the possibility to evaluate accessibility values in relation to the population affected, allowing a better assessment of the overall accessibility situation.

Method of raster-based accessibility analysis

Based on the coordinates of the raster centroids of the AIT-GRID, the road network and the coordinates of the street petrol stations for every raster centroid the shortest street distance to the next street petrol station has been calculated. Thus, the accessibility analysis is based implicitly on the assumption that from every start-point the next street petrol station is chosen.

Because of the huge amount of data (street network covering the whole of Germany) it was decided to use PostgreSQL/PostGIS for data preparation prior to the actual accessibility analysis. Neverthe-

[21] This represents a subset of the AIT-GRID for Germany as cells split by clipping methods have been counted as one cell and cells that could not be attributed to a BBSR-Kreistypen 2009, federal state or community have not been included in the analysis. (This affected cells at the border of the data set where, because of minor geographical inconsistencies in the different data sets available the data sets differ from one another due to their geographical accuracy/coverage).

less, although PostgreSQL/PostGIS can be extended with network analysis functions with the „pgdijkstra library" it was decided to perform the shortest path calculation based on the „Dijkstra shortest path" algorithm (Dijkstra 1959) with the help of the Perl-Module „Graph-0.94". Two reasons were decisive: First, the hardware available for the analysis proved to run out of resources when trying the perform the whole analysis within PostgreSQL/PostGIS. Second, we wanted to be able to write every completed shortest path calculation to file/database, so that in the case of an error, power shortage, etc. the computation process could easily be resumed with the remaining starting points. Altogether the workflow for the accessibility analysis contained following steps:

1. Building of a network topology
    (a) Determination of the road/line next to every start/target coordinate;
    (b) Determination of the point where the lead from every start/target point crosses the road/line next to the point;
    (c) Adding of a new line to the road network from the start/target point to the point determined by c;
    (d) Splitting of the original line/road in the network at the point determined by c;
    (e) Building of a topological network graph for the resulting street data set by defining the start and end point of every line segment and allocating unique node id's to the defined start and end points;
    (f) Allocation of network node id's to the corresponding start and target points.
2. Shortest path analysis
    (a) Due to performance issues it was decided to perform a catchment area analysis prior to the shortest-path analysis. On the one hand, the three petrol stations nearest to the starting-point under consideration were first defined based on Euclidean distance, and on the other hand the route network to be considered by the shortest path algorithm was limited to only the subset of streets which are within the distance to the point furthest away from the starting-point plus 1,000 meters (e.g., Figure 2). This allowed the potential 2,974,239,051[22] routing requests to be reduced to 627,609[23] requests and the computation time necessary to perform the whole accessibility analysis to be reduced quite considerably[24].
    (b) Determination of the shortest distance (in meters) between start point and every potential target point based on the „Dijkstra shortest path" algorithm;
    (c) Allocation of the shortest paths to the raster cell of the AIT-Grid.

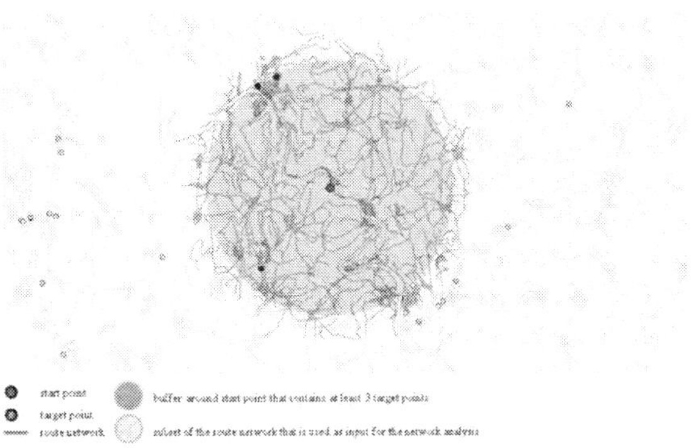

Figure 2: Method of catchment area analysis prior to the routing analysis.

6. Accessibility of Street Petrol Stations – comparing results based on ESM with results based on OSM

After having introduced the data sets as well as the methodology upon which the accessibility analysis is based, I now want to present some of my findings on street petrol station accessibility in Germany. Thereby two aims are pursued: (1) The assessment of the street petrol station accessibility situation in Germany and (2), a comparison of the performance of ESM and OSM in order to approach the question of the usability of OSM within research focusing on rural studies. Due to the accessibility results it has to be kept in mind that accessibility is not static but changes during the course of time. In this sense, the results presented represent a snap-shot for the reference years 2011/2012. Furthermore the results are strongly dependent on the accessibility model used; they do not represent reality but, as common

[22] 209.203*14.217 (start points * target points).
[23] 209.203*3 (start points * next three target points).
[24] Three weeks instead of more than 95 years (estimated based on computation time of single queries).

with models, an ideal typical situation (Johnston et al, 2000, 508 ff.).

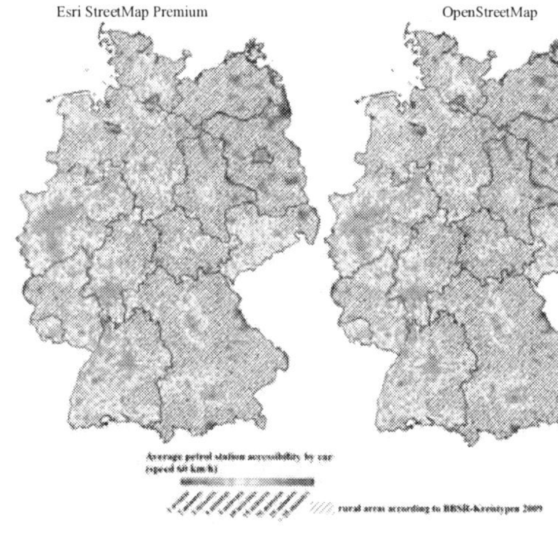

Figure 3: Street petrol station accessibility by car (speed 60 km/h) from ESM data (left) and OSM data (right) (Sources: Administrative boundaries: Bundesamt für Kartographie und Geodäsie 2010; BBSR-Kreistypen 2009: BBSR; Raster: © EFGS, 2009).

Only petrol stations within Germany were included in the accessibility analysis. Petrol stations in neighboring foreign countries were not considered. Therefore in areas close to the country border the model results are likely to diverge more from reality than in regions at greater distances to the country borders. In Figure 3 the accessibility of street petrol stations by car (speed 60 km/h) according to the ESM is contrasted to that acquired with the help of the OSM. Visually, by and large both results are equal to each other. But what do the actual data sets say that is revealed by the map? Does the street network used as a basis for the accessibility analysis make a difference concerning the results obtained? In order to answer this question we have to compare the results with each other. The results of the two accessibility analyses represent related samples as we computed both based on the same starting-points. Furthermore the resulting values proved not to be normally distributed. Therefore it was decided to execute the Wilcoxon-test, a non-parametric statistical hypothesis test, in order to determine if the results differ statistically from one another. The two-sided Wilcoxon-test [25] performed with a confidence interval of 0.95 delivered a V of 9395891083 and a p-value 2.2e-16. So as p 0.025 it can be reasoned that the results gained by the ESP dataset differ from that gained by the OSM in a statistical sense. But what does this mean in practical terms? In order to approach this question we will have a look at the accessibility results computed based on ESP as well as OSM, first.

In Germany the nearest street petrol station is accessible in 5.4 km on average according to ESP and 5.5 km according to OSM (see Table 2).

Considered according to the BBSR-Kreistypen 2009 it can be observed that the average distance to be covered to reach the next street petrol station increases from „Core cities in agglomerations" (ESP: 1.9 km; OSM: 2.0 km) to „Sparsely populated rural areas" (ESP: 7.8 km; OSM: 7.9 km). This observation suggests that a correlation exists between BBSR-Kreistyp 2009 and distance to be covered to reach the next petrol station. The existence of such a correlation is statistically proven by the measure of association η [26], which amounts to 0.4 for the ESM as well as for the OSM data and indicates the existence of a medium statistical correlation. All in all it can be concluded that with increasing peripherality, the distance to be covered to reach the next petrol station in Germany increases, too.

Similar to the BBSR-Kreistypen 2009 the average distances to be covered to reach the next street petrol station differ within the single federal states between Berlin (ESP/OSM: 2 km) and Mecklenburg-Vorpommern (ESP/OSM: 8 km) (see Table 2).

Here, too, a possible statistical correlation between distance to be covered and federal state can be determined by considering the measure of association η, which amounts 0.3 for the ESM as well as OSM data and indicates the existence of a medium statistical correlation. All in all, similar to the BBSR-Kreistypen 2009, it can be concluded that the federal state in which a person lives influences the distance to be covered to reach the next petrol station in Germany. Comparing the average accessibility values for the statistical units „Germany", „BBSR-Kreistypen 2009" and federal states, the difference between the values based on ESP and those based on OSM are considered less than 150 m in all regions

[25] The Wilcoxon-test was performed with the statistics program R (p: probability value; R's V corresponds the Wilcoxon W that is the sum of positive and negative ranks).

[26] η (Eta) is a measurement of association that enables to compute the statistical correlation between an independent nominal variable and a metric dependent variable. The value range is between 0 and 1 whereas η=0 means that no correlation exists. As a rule of thumb η can be interpreted as follows: $0<\eta<0.2$: extremely weak correlation; $0.2<\eta<0.4$: weak correlation; $0.4<\eta<0.6$ medium correlation; $0.6<\eta<0.8$: strong correlation; $0.8<\eta<1$: very strong correlation; η=1: perfect correlation. The measurement of association η does not allow to determine the direction of the correlation.

(see Table 2). So, considering the accessibility values aggregated for the BBSR-Kreistypen 2009, as well as the federal states, no noteworthy differences can be identified between the results obtained based on ESM or OSM.

Table 2: Comparison of average street petrol station accessibility for selected statistical units based on ESP data set compared to Open StreetMap data set (Source: Own calculation).

statistical units	Esri StreetMap Premium						Open StreetMap						Difference (OSM-ESM)			
	stdev.	var.	meters				stdev.	var.	meters				meters			
			m	avg.	min.	max.			m	avg.	min.	max.	m	avg.	min.	max.
Germany	3772.64	61.42	4637.91	5394.12	18.06	59169.00	3850.18	62.05	4693.65	5466.32	19.54	51569.00	4.76	72.20	-42516.21	41396.30
BBSR-Kreistypen 2009																
1 Core cities in agglomerations	1378.13	37.12	1563.76	1899.02	30.78	12631.98	1484.37	38.53	1570.77	1953.26	35.01	14805.26	0.69	54.238	-6945.61	7570.86
2 Densely populated distcicts in agglomerations	2075.27	45.56	2725.93	3159.54	19.60	18007.39	2155.58	46.43	2774.53	3231.46	19.54	23637.26	4.11	71.922	-15276.83	11485.88
3 Highly populated districts in agglomerations	2970.48	54.50	4114.43	4639.36	29.46	34251.03	3006.94	54.84	4180.20	4696.68	28.17	23543.51	5.15	57.319	-26516.17	16244.06
4 Rural districts in agglomerations	4311.71	65.66	6162.09	6828.79	111.16	59169.00	4357.36	66.01	6191.18	6896.80	110.33	51569.20	3.94	68.009	-42516.21	26675.94
5 Core cities in urbanised areas	1797.60	42.40	2093.92	2522.40	22.51	13943.58	1901.13	43.60	2126.83	2579.00	22.18	23205.99	1.70	56.604	-5650.77	20281.78
6 Densely populated areas in urbanised areas	2907.19	53.92	4148.79	4650.75	18.06	31460.88	3025.52	55.00	4196.99	4719.52	32.28	42198.00	4.70	68.768	-17134.09	31348.41
7 Rural districts in urbanised areas	3716.53	60.96	5722.11	6217.40	27.42	40980.98	3859.70	62.13	5792.24	6340.81	25.54	46953.76	8.38	123.41	-24421.86	41396.30
8 Densely populated rural areas	3519.52	59.33	5369.68	5868.62	27.33	54141.78	3570.80	59.76	5407.01	5913.90	28.10	47972.24	4.61	45.276	-33265.31	34186.50
9 Sparsely populated rural areas	4889.49	69.92	7053.66	7811.93	39.07	48879.79	4939.45	70.28	7092.72	7863.12	33.48	44368.30	3.53	51.197	-26359.42	24496.69
Federal states																
Schleswig-Holstein	3472.45	58.93	4778.67	5365.07	36.88	33634.92	3390.72	58.23	4821.67	5350.77	36.88	32132.01	1.58	-14.29	-19293.86	26675.94
Hamburg	1553.69	39.42	1436.14	1955.31	59.76	9993.74	1676.56	40.95	1504.51	2037.02	55.16	9834.29	1.80	81.71	-4886.56	5479.56
Niedersachsen	3151.31	56.14	4695.43	5134.94	37.17	34251.03	3213.79	56.69	4718.59	5166.57	39.37	42198.00	1.68	31.63	-26516.17	31348.41
Bremen	1603.93	40.05	1599.32	2039.44	116.73	10090.76	1614.48	40.18	1618.67	2070.27	75.52	11454.69	0.69	30.83	-4119.07	4683.78
Nordrhein-Westfalen	2131.25	46.17	2930.77	3341.34	26.99	19478.35	2194.90	46.85	2990.91	3410.53	34.32	23205.99	3.92	69.19	-11694.94	20281.78
Hessen	2883.67	53.70	3624.75	4227.78	19.60	17754.22	3015.69	54.92	3690.49	4309.20	19.54	21991.40	4.35	81.42	-15276.83	16118.26
Rheinland-Pfalz	3807.99	61.71	4899.88	5677.62	22.51	28155.06	3996.36	63.22	4947.42	5781.62	22.18	30710.24	9.68	104.00	-14676.53	14600.52
Baden-Württemberg	3442.84	58.68	4173.37	4911.94	21.29	40980.98	3539.88	59.50	4270.08	5001.60	25.54	38568.59	9.16	89.67	-24421.86	23592.70
Bayern	3497.67	59.14	4992.52	5545.26	18.06	54141.78	3548.47	59.57	5016.87	5591.63	28.10	32617.54	5.85	46.37	-33265.31	20889.38
Saarland	3251.03	57.02	3410.23	4264.37	122.16	20165.05	3404.42	58.35	3495.18	4393.56	114.51	23543.51	6.79	129.19	-5233.76	7422.76
Berlin	1288.63	35.90	1387.31	1707.66	30.78	12631.98	1299.30	36.05	1322.94	1655.72	35.01	14805.26	-4.41	-51.94	-6945.61	7541.02
Brandenburg	5065.29	71.17	7173.29	7954.05	111.16	59169.00	5186.06	72.01	7273.02	8097.70	110.33	51569.20	6.10	143.65	-42516.21	41396.30
Mecklenburg-Vorpommern	5008.15	70.77	7564.53	8118.41	55.70	48879.79	4987.06	70.62	7610.80	8181.88	46.18	44368.30	3.96	63.47	-26359.42	24496.69
Sachsen	3568.31	59.74	4854.92	5490.21	61.58	43971.34	3713.20	60.94	4973.87	5634.56	61.75	32290.57	8.00	144.35	-12857.06	11330.54
Sachsen-Anhalt	4187.20	64.71	6305.46	6888.08	50.91	28473.93	4326.43	65.78	6361.00	6981.82	50.91	47972.24	5.84	93.75	-15176.64	34186.50
Thüringen	3694.95	60.79	5710.73	6197.92	74.62	23420.83	3850.81	62.05	5835.30	6333.63	69.73	24879.75	9.49	135.71	-11385.68	13863.07

Table 3: Difference of street petrol station accessibility between ESM and OSM.

spatial category	nr of cells	distance in m				
		<= 150	> 150 to 300	> 300 to 500	> 500 to 1000	> 1000
		% cells per cells in category				
all cells	209,203	59.6	10.8	8.1	9.7	11.7
rural areas	112,357	58.0	10.6	8.1	9.9	13.4
urban areas	96,846	61.4	11.2	8.2	9.5	9.7
1 Core cities in agglomerations	7,863	65.9	11.7	8.4	8.1	5.9
2 Densely populated distcicts in agglomerations	19,678	60.6	11.9	8.8	10.1	8.6
3 Highly populated districts in agglomerations	18,618	60.2	11.0	8.4	9.5	10.9
4 Rural districts in agglomerations	13,956	59.1	10.7	7.7	9.2	13.3
5 Core cities in urbanised areas	3,484	64.5	10.6	8.1	8.6	8.2
6 Densely populated areas in urbanised areas	47,203	61.3	10.9	7.9	9.6	10.4
7 Rural districts in urbanised areas	40,904	55.6	11.0	8.9	10.8	13.8
8 Densely populated rural areas	31,963	60.5	10.5	7.6	9.6	11.7
9 Sparsely populated rural areas	25,534	58.3	9.9	7.5	9.4	15.0

Nevertheless comparing the average distances at the community level it can be observed that the distances acquired by ESM differ quite considerably from those acquired by OSM. For 47 % of Germany's 11,589 communities, the difference is greater than 150 meters, and for 27 % of the communities greater than 300 meters.

Thereby 72% of the communities with differences of 150 meters belong to the BBSR-Kreistypen 2009 „rural districts in urbanised areas" (1516), „densely populated areas in urbanised areas" (1345), „densely populated rural areas" (763) and „sparsely populated rural areas" (727). Last but not least, a comparison based on the single raster cells of the AIT-GRID reveals that the computed accessibility values differ, sometimes quite considerably, from one another. In 69% of the rural cells (compare Table 3) the difference between ESM and OSM (absolute value) is less than 300 m (a value, we think is still acceptable when considering driving distances, but not when considering walking distances). Surprisingly, urban areas do not differ much from rural areas as a whole.

Also, considering differences greater than 1,000 m, no noteworthy rural (13 % of the „rural" raster cells) or urban (10 % of the „urban" raster cells) differences can be found.

Nevertheless, as can be seen in Figure 4 as well, the cells/regions with the greatest differences can be found in the federal states Mecklenburg-Vorpommern, Brandenburg, Thüringen and Sachsen-Anhalt, northern Bayern, western Rheinland-Pfalz, eastern Niedersachsen and the Alps.

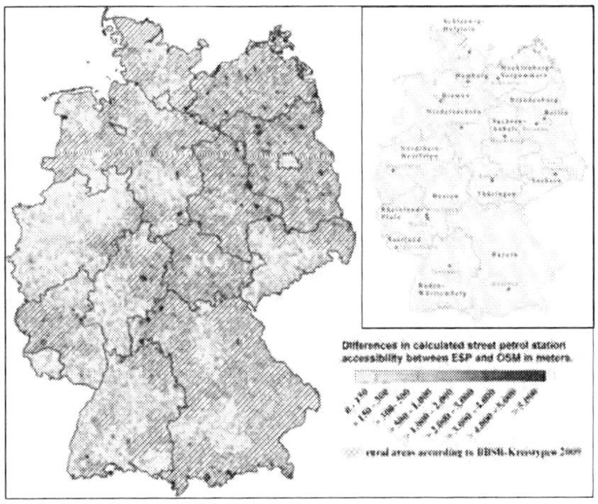

Figure 4: Differences in street petrol station accessibility between ESM and OSM. (Administrative boundaries: Bundesamt für Kartographie und Geodäsie 2010; BBSR-Kreistypen 2009: BBSR; Raster: © EFGS, 2009).

Population and Street Petrol Station accessibility

The identification of the areas with the greatest distances to be covered present a first indication for areas with potential deficits with regard to accessibility, respectively supply shortage, but they do not allow the population affected to be inferred . But this is important for an assessment of the accessibility situation, as a below average accessibility may be rated differently in regions with a high share of population than in regions with a low share of population. Such a consideration is enabled by merging the accessibility results with the population data of the AIT-GRID. Against the background that petrol stations are operated under economic aspects it is not surprising that a modest negative correlation[27] can be found between the accessibility of petrol stations and population (ESM/OSM r of -0.32). By trend, the longer the distances to be covered to reach the next petrol station are,the lower the population value is within a raster cell. This finding suggests that especially regions with low population are affected by disadvantageous street petrol station accessibility.

Table 4: Accessibility of street petrol stations according to BBSR-Kreistypen 2009 and population per BBSR-Kreistyp 2009 according to the AIT-GRID (Source: Own calculation).

BBSR-Kreistyp 2009	population per BBSR-Kreistyp 2009 in % according to the AIT-Population Density Grid							
	<=5,400 m		> 5,400 m <=10,000 m		> 10,000 m <= 15,000 m		> 15,000 m	
	ESM	OSM	ESM	OSM	ESM	OSM	ESM	OSM
1 Core cities in agglomerations	99.54	99.44	0.43	0.52	0.03	0.04	0.00	0.00
2 Densely populated distcicts in agglomerations	96.49	96.03	3.33	3.78	0.18	0.19	0.00	0.01
3 Highly populated districts in agglomerations	86.36	85.63	11.75	12.22	1.64	1.91	0.25	0.24
4 Rural districts in agglomerations	73.51	73.30	18.34	18.36	6.73	6.62	1.42	1.72
5 Core cities in urbanised areas	98.53	98.24	1.40	1.70	0.07	0.06	0.00	0.00
6 Densely populated areas in urbanised areas	84.78	84.04	13.12	13.72	1.91	2.02	0.18	0.21
7 Rural districts in urbanised areas	69.36	68.43	23.05	23.06	6.48	7.16	1.12	1.35
8 Densely populated rural areas	76.12	75.64	18.36	18.46	4.97	5.23	0.55	0.67
9 Sparsely populated rural areas	64.04	63.97	21.83	21.63	10.31	10.38	3.82	4.02

All in all the portion of the population that is able to reach a street petrol station by car (60 km/h) within 5.4 minutes (average time/distance according to ESM) is based on the ESP data 87.43 %, based on the OSM data 86.93 % Within 15 minutes[28] according to the ESM data set 99.56 % and according to the OSM data set 99.50 % of the German population is able to reach the next petrol station by car (60 km/h). Comparing the street petrol station accessibility within the BBSR-Kreistypen 2009 (Table 4) it becomes obvious that in rural regions the share of population that has to cover greater distances to reach the next street petrol station is greater than in urban regions.

But all in all, in all BBSR-Kreistypen 2009, except in sparsely populated rural areas (96 % ESM/OSM), over 98 % (ESM/OSM) of the population can reach the next street petrol station by car (60 km/h) within 15 minutes. In summary, comparing the average accessibility of street petrol stations in Germany according to the BBSR-Kreistypen 2009 no noteworthy differences or deficits can be found.

Synthesis – Street Petrol Station accessibility in Germany

The results on street petrol station accessibility in Germany as key service supporting the mobility of citizens in rural areas can be summarized as follows:

- In Germany, a street petrol station can on average be reached within 5.4 km (ESM)/ 5.5 km (OSM);
- The higher the absolute population within a region, the shorter the distances to the next street petrol station;
- With increasing peripherality, the distance to the next street petrol station increases;
- The distance to the next street petrol station is dependent on one's location within the federal states, whereby in tendency the distances are longer in the „new" federal states[29] than in the „old" federal states[30];
- At least 87.4 % (ESM)/ 86.9 % (OSM) of the German population is able to reach the next street petrol station within 5.4 minutes by car (60 km/h), and another 12.1 % (ESM)/12.5 (OSM) within 15 minutes.

Altogether these findings suggest that in Germany, despite of the decreasing number of street petrol stations since 1970 and the fact that Germany belongs to the European countries with the lowest density of petrol stations, street petrol stations are for the majority of the population, quite accessible. Regions with disadvantageous street petrol station accessibility are

[27] Pearsons's product moment correlation coefficient..

[28] This is a time-span that according to a study of the „Amt für Raumentwicklung und Geoinformation" of the canton St. Gallen, Switzerland (2008) is commonly accepted to reach SGI.

[29] Former German Democratic Republic.

[30] Federal States of the Federal Republic of Germany before the reunification with the German Democratic Republic.

predominantly sparsely populated. These findings show that, based on the model results, no need for action or intervention can be identified.

7. Lessons learned

The following résumé summarizes the experiences as well as first subjective impressions due to the usability of the data sets as well as the software used. Following this consideration we will conclude with a consideration of the usability of OSM data compared to commercial routing networks within rural studies.

ESRI StreetMap Premium

The commercial ESP dataset proved to be quite comprehensible. The dataset available (Esri StreetMap Premium for mobile) contained streets as lines as well as street crossings as points allowing to differentiate between street crossings and bridges/tunnels/dead-ends. Unfortunately, in order to obtain a topological street network the street crossings had to be projected into the street-lines dataset first. Subjected to the size of the street network under consideration this process can consume quite a lot of computation time. Nevertheless the process could be performed without greater difficulties. Minor topology inconsistencies like unconnected lines and street crossings lead to a number of 1,484 (ca. 0.7 %) start points for which no routing could be performed. For these points the simple Euclidean distance had been determined in order to compensate the missing values.

OpenStreetMap

Compared to the ESP dataset the OSM dataset, proved to be less comprehensible and required more time to figure out its internal structure and how to obtain the data necessary for building a street network fit for performing network analyses. Especially the presentation and description of the OSM in the form of a „wiki"makes it, according to my perception, quite difficult to become quickly acquainted with its structure. The availability of a concise and printable document describing how to get the data step-by-step and what to do in order to use it would have been very helpful – especially if one only wants to use the data without getting too involved in all the technical detail necessary to participate in the OSM project.

One major drawback experienced was the fact, that after downloading and importing the route data in the database, because of topological inconsistencies/errors, the dataset proved not to be suitable for routing purposes. That means lines obviously connected in reality are split because of dangling nodes, one aspect we notice because only for 2,800 of the 209,203 starting points were routes to targets (petrol stations) identifiable using the OSM obtained out of the box. This behaviour could be corrected by connecting all dangling nodes in the network graph within a distance of 1 m, but at the cost of keeping information about dead ends between the nodes affected.

Another major drawback is the seemingly missing or insufficient standardization, respectively quality check, concerning the tags used for identifying/differentiating feature from one another[31]. This makes it quite difficult to determine all the lines to be included in the street network, always leaving some kind of uncertainty if something has been left out or added, and which may in consequence distort the network under consideration. In consequence, one gets the impression that the resulting street dataset represents a less exact model of reality than that based on the Esri StreetMap Premium data set.

Because of the difficulties in easily getting concise information about the data set and its use, but especially because of the fact that practically everybody is able to edit or change the dataset and that the dataset contains different regional states of data acquisition – raising questions about its reliability – using the data in scientific analysis causes some uneasiness due to the reliability of the analysis results.

Similar to the ESM dataset, because of topology inconsistencies that could not easily be compensated for 2,441 of the starting points under consideration (ca. 1.2 %), no routing could be performed based on the network. For these points the simple Euclidean distance was determined in order to compensate the missing values.

PostgreSQL/PostGIS

PostgreSQL with the PostGIS extension proved to be a very valuable tool for data preparation especially when huge amounts of data are to be analysed and edited. In contrast to other non open source GIS tools, PostgreSQL had no problems in handling quite

[31] A good example is here the tag for the amenity „bank" (financial institute) that is sometimes mixed up with the amenity „park bench" which is, at least in the German OSM dataset sometimes also tagged as „bank".

huge data sets. Nevertheless, dependent on the GIS-functions utilized, computing times for completing queries can be quite long (hours/days). For example it took approximately 12 hours to connect the raster centroids as well as petrol stations with the road network.

The accessibility calculation took approximately two weeks. So in order to effectively handle and analyse large sets of data it proved to be important to use indices as often as possible as this speeds up the data processing within PostgreSQL quite considerably. Provided that database, SQL and GIS basics are known, the specific PostGIS syntax for performing GIS analyses can be quite easily learned (cp. `http://trac.osgeo.org/postgis/`.(13.05.2013)). Nevertheless random-access-memory (RAM) proved to be a limiting factor when using PostgreSQL with huge datasets and complex long-running queries. So it was not possible to perform the actual shortest path analysis with PostgreSQL tools alone.

Shortest path analysis

Although PostgreSQL/PostGIS can be extended with shortest-path routing functionalities with the pgRouting extension, it was decided to perform the routing outside the database via the Perl-Module „Graph-0.94. This proved to be necessary as calculating the shortest path for every start-point proved to be so memory-intensive that the analysis failed to complete within PostgreSQL and the hardware available because of memory shortage.

Therefore an external Perl-Script has been developed that is able to read the source, target and route information from the database, perform the catchment as well as network analysis and write back the analysis result to the database. Thereby, the Perl-Module „Graph-0.94" turned out to be quite easy to use and offer a range of interesting graph analysis functions. One advantage of performing the routing via an external Perl script was the ability to use parallel processing abilities which proved to be quite handy when analysing, for example, different categories of petrol stations. Furthermore status messages could easily be implemented allowing the progress of the analysis to be followed.

Performance of OSM compared to ESM

The direct comparison of the accessibility values calculated based on ESM with those calculated based on OSM showed that the resulting values differ, partly considerably, from one another. Nevertheless, some kind of a pattern seems to exist as the regions/cells with the greatest distance differences can be found in the federal states Mecklenburg-Vorpommern, Brandenburg, Thüringen and Sachsen-Anhalt, northern Bayern, western Rheinland-Pfalz, eastern Niedersachsen and the Alps. What does this all say about the usability of OSM data for rural studies?

Considering the findings of Ludwig et al (2010), as well as Zielstra and Zipf (2010), due to OSM completeness as well as the fact that tags for identifying features within the OSM dataset (e.g., residential streets, highways, etc.) are not really checked for quality, and the fact that the data set has to be topologically „corrected" in order to be able to build a graph capable for routing at the cost of topological information (dead ends, bridges, tunnels) it can be assumed that the OSM still represents a less accurate model of reality than the ESM.

Therefore it can be assumed that the distances calculated based on OSM are less accurate than those based on commercial routing networks like for example ESM.

Considering the findings of Ludwig et al. (2010), as well as Zielstra and Zipf (2010) due to OSM completeness it is surprising that the accessibility values differ within urban as well as rural areas within the same range.

This finding was unexpected because we assumed that a clear distinction between urban and rural areas would exist, with greater differences in the accessibility values computed in rural areas. Nevertheless although great differences between the distances calculated could be found on a per cell comparison as well as on community level, aggregated average accessibility values obtained by ESM as well as OSM for greater aggregates like country, federal states or BBSR-Kreistypen 2009 proved to be comparable to each other.

The shortcomings of the OSM – especially the uncertainty about accuracy and completeness – together with the comparatively high learning level necessary to be able to use the OSM makes it, in my opinion, a less useful product within scientific studies focusing especially on rural areas, especially for small-scale analyses at the county level and below or the deduction of a small-scale accessibility indicator. But if one is not interested in the single cell values or small-scale analyses, the OSM-data proved to perform similarly to that of commercial alternatives. In such cases, the results suggest that the OSM may indeed provide an interesting low/no cost alternative.

References

ADAC (2012), http://www.adac.de/infotestrat/tanken-kraftstoffe-und-antrieb/probleme-tankstelle/anzahl-tankstellen-markenverteilung/ (11.09.2012).

Amt für Raumentwicklung und Geoinformation, Kanton St. Gallen (2008), 'Erreichbarkeit Grundversorgung', http://www.sg.ch/home/_bauen__raum___umwelt/raumentwicklung/raumbeobachtung/indikatoren_zur_raeumlichen.____etc____medialib____dokument_library____bauen__raum___umwelt____raumentwicklung____raumbeobachtung.Par.0013.FileRef.tmp/ErreichbarkeitGrundversorgung_Kanton.pdf (24.03.2011).

BAKBASEL (2009), 'Erreichbarkeitsanalysen', http://www.bakbasel.ch/downloads/competences/location_factors/accessibility/bakbasel_factsheet_accessibility_dt.pdf

Barlösius, E.; Neu, C. (Ed.) (2008), 'Peripherisierung – eine neue Form sozialer Ungleichheit?', Berlin-Brandenburgische Akademie der Wissenschaften. Interdisziplinäre Arbeitsgruppe Zukunftsorientierte Nutzung ländlicher Räume Materialien Nr. 21.

Bleisch, A. (2005), 'Die Erreichbarkeit von Regionen. Ein Benchmarking-Modell', Dissertation an der wirtschaftswissenschaftlichen Fakultät der Universität Basel.

Bundesministerium für Verkehr, Bau und Stadtentwicklung (2009), http://www.mobilitaet-indeutschland.de/pdf/MiD2008_Praesentation_Abschlussver-anstaltung_August2009_FassungMaerz2010.pdf (19.04.2011).

Bundesministerium für Verkehr, Bau und Stadtentwicklung (2009a), http://www.bbsr.bund.de/nn_1067638/BBSR/DE/Raumbeobachtung/Raumabgrenzungen/SiedlungsstrukturelleGebietstypen/Kreistypen/kreistypen__node.html?__nnn=true (30.10.2012).

Bundesministerium für Verkehr, Bau und Stadtentwicklung (2011), 'Raumordnungsbericht 2011', Deutscher Bundestag. Drucksache 17/8360. 17. Wahlperiode. 13.01.2012.

Bleisch, A.; Koellreuter, C. (2003), 'Die Erreichbarkeit von Regionen' http://e-collection.ethbib.ethz.ch/eserv/eth:27008/eth-27008-01.pdf.

Dahlgren, A. (2008), 'Geographic Accessibility Analysis - Methods and Application', Dissertation Lund Institute of Technology. http://www.dissertations.se/dissertation/ff24596a13/ (18.10.2011).

Dewenter, R.; Haucap, J.; Heimeshoff, U. (2012), 'Maßnahmen zur Steigerung des Wettbewerbs auf den Kraftstoffmärkten in Deutschland. Studie für den Allgemeinen Deutschen Automobilclub e.V. (ADAC)',

Dijkstra, E. W. (1959), 'A Note on Two Problems in Connexion with Graphs', Numerische Mathematik, 1, 269-271.

Gilgen, k.; Sartoris, A. (2005), 'Grundversorgung mit Gütern des täglichen Bedarfs. Eine Methode zur Analyse der Grundversorgung anhand eines Testes im Bleniotal und in der Leventina (kt. Tessin). Das Grundversorgungskonzept und dessen Umsetzung im Rahmen einer Grundversorgungspolitik', Institut für Raumentwicklung, Hochsschule für Technik, Rapperswil, http://www.irap.ch/uploads/tx_hsrpm/Schlussbericht_Grundversorgung_2005_mit_Anhang_02.pdf (24.03.2011).

Gyllensvärd, U. (1999), 'Der Convenience-Handel auf dem Weg in das nächste Jahrtausend', in: Tomczak, T. (Hrsg.): Alternative Vertriebswege. – Stuttgart: Schäffer-Poeschel, 184-193.

Hemetsberger, M.; Ortner, S. (2008), 'Erreichbarkeitsbasierte Raster Raumanalyse – Anwendungen in der Landesplanung', in: Schrenk, M., Popovich, V., Engelke, D., Elisei, P. (Eds..): REAL CORP 008 Proceedings/ Tagungsband, 255-263.

Heinze, G.W.; Herbst, D. & Schühle, U. (1982), 'Verkehrsnachfrage und Verkehrsangebot im ländlichen Raum' Der Landkreis: Zeitschrift für kommunale Selbst-verwaltung 52/8-9; 367-370, Ed.: Der Deutsche Landkreistag, Kohlhammer Verlag; Stuttgart.

Johnston, R. J.; Gregory, D.; Pratt, G.; Watts, M. (2000): 'The Dictionary Of Human Geography', 4th. Ed. Blackwell.

Korn, J. (2006): 'Transiträume als Orte des Konsums - eine Analyse des Standorttyps unter besonderer Berücksichtigung der Bahnhöfe', Dissertation im Fach Geographie an der Mathematisch-Naturwissenschaftlichen Fakultät II der Humboldt-Universität zu Berlin.

Kulke, E. (1994): ',Tante-Emma' in neuem Kleid? Zur Funktion von Tankstellenshops und Kiosken im Versorgungsnetz des Einzelhandels', Die Erde, Nr. 123, 181-196.

Küpper, P.; Steinrück, B. (2010), 'Mobilität in ländlichen räumen unter besonderer Berücksichtigung bedarfsgesteuerter Bedienformen des ÖPNV' (= Arbeitsberichte aus der vTI-Agrarökonomie 02/2010), http://www.vti.bund.de/de/institute/lr/publikationen/bereich/ab_02_2010_de.pdf (15.10.2012).

Leser, H.; Haas, H.-D.; Moismann, T.; Paesler, R. (1993), 'Diercke. Wörterbuch der Allgemeinen Geographie', Band 1 A-M. Braunschweig.

Ludwig, I.; Voss, A.; Krause-Traudes, M. (2010), 'Wie gut ist Open street Map? Zur Methodik eines automatisierten objektbasierten Vergleichs der Strassennetze von OSM und NAVTEQ in Deutschland',gis.science. Die Zeitschrift für Geoinformatik, 32, 4, 148-158.

Meyer, W. (2006), 'Zeige mir die Achse des Bösen – Geostatistische Welt-Bilder',in: Strobl, J.; Blaschke, T.; Griesebner, G. (Ed.): Angewandte Geoinformatik 2006. Beiträge zum 18. AGIT-Symposium, Salzburg, Heidelberg, 454-459.

Maier, J; Forster, S.; Frankenberger, M. Hofman, T. (2010), 'Regional-Atlas Wirtschaftsband A9 Fränkische Schweiz', http://www.goessweinstein.de/fileadmin/Gemeinde/Informationen/Nachrichten_anderer_Stellen/Wirtschaftsband_A9/End-Version_Regional-Atlas_WBA9FS_Stand_28.12.2010.pdf (23.03.2011).

Morgenroth, O. (2008), 'Zeit und Handeln. Psychologie der Zeitbewältigung', Stuttgart.

Morgenstern, T.; Zimmermann, K. (2011), 'Branchenstudie Tankstellenmarkt', URL: http://www.bft.de/bft/download/btsmarkt11.pdf (11.09.2012).

Neumeier, S. (2013), 'Modellierung der Erreichbarkeit von Straßentankstellen. Untersuchung zum regionalen Versorgungsgrad mit Dienstleistungen der Grundversorgung', in preparation.

OECD (2007), 'OECD-Prüfbericht zur Politik für ländliche Räume', Deutschland.

Platzer, G.; Gmeinhart, G. (2003), 'Modellierung von Erreichbarkeit im Öffentlichen Verkehr mittels GIS Gesamtlösung zur Bewertung und Analyse der Angebotsqualität im Öffentlichen Verkehr', in: „CORP 2003" Computergestützte Raumplanung Beiträge zum 8. Symposium zur Rolle der Informationstechnologie in der und für die Raumplanung, Wien, 2003.

Reudenbach C., Schulze U. (2010), GIS für Geographen - eine Einführung. http://gisbsc.gis-ma.org/GISBScL2/de/text/GISBScL2.pdf. (23.08.2011)

Schulz, A.-C.; Bröcker, J. (2007), 'Die Erreichbarkeit der Arbeitsmärkte für Berufspendler aus den Gemeinden Schleswig-Holsteins', in: IAB regional, 1.

Schürmann, C.; Spiekermann, K.; Wegener, M. (1997), 'Accessibility indicators', Berichte aus dem Institut für Raumplanung 39, Institut für Raumplanung, Dortmund.

Schürmann, C. (2008), 'Berechnung verschiedener Erreichbarkeitsindikatoren für den Ostseeraum', http://www.brrg.de/_content/documents/publications/dak08_erreichbarkeit.pdf.

Schwarze, B. (2005), 'Erreichbarkeitsindikatoren in der Nahverkehrsplanung', Institut für Raumplanung, Universität Dortmund - Fakultät Raumplanung, Arbeitspapier 184.

Spiekermann, K.; Wegener, M. (2008), 'Modelle in der Raumplanung I. 5 Erreichbarkeit I', http://www.spiekermann-wegener.de/mir/pdf/MIR1_5_181108.pdf.

Steinocher, K.; Köstl, M., Weichselbaum, J. (2011), 'Grid-based population and land take trend indicators – new approaches introduced by the geoland2 core Information Service for Spatial Planning', NTTS conference, Brussels, Feb 2011, CD-ROM.

Streit, U. (2011), http://ifgivor.uni-muenster.de/vorlesungen/Geoinformatik/kap/kap4/k04_6.htm (18.04.2011).

Winkel, R.; Greiving, S. Pietschman, H (2007), 'Sicherung der Daseinsvorsorge und Zentrale-Orte-Konzepte – gesellschaftspolitische Ziele und räumliche Organisation in der Diskussion. Stand der Fachdiskussion', BMVBS-Online-Publikation12/2010, http://www.bbsr.bund.de/cln_016/nn_21918/BBSR/DE/Veroeffentlichungen/BMVBS/Online/2010/ON122010.html (24.03.2011).

Zielstra, A.; Zipf A. (2010), 'A comparative study of proprietary geodata and volunteered Geographic Information for Germany', AGILE 2010, The 13th AGILE International Conference on Geographic Information Science, Guimarães.

Bytecode unification of geospatial computable models

by Jan Kolář

Grifinor Project. jan.kolar@grifinor.net

Abstract

Geospatial modelling revolves around the structures of data and the semantics of these structures. This is enough in simple cases, but becomes insufficient when the best structure and semantics is hard to find or the solution is too heterogeneous to fix and reuse. Field-based and objects-based geospatial models often share common GIS data structures interchangeably, but their all possible meanings are too many to define in an immutable manner. Less studied approach to geospatial modelling is using mutable structural properties and their semantic interpretation. This work shows that the functional aspect of geospatial models is just as important as the structural and semantic aspects. It also shows that semantic and even structural properties may change when functionality is integral part of the data model, and not exclusively separated at software implementation level. The paper uses this modelling paradigm to address the divide caused by field-based and object-based data models, and other challenges regarding synergy of geospatial systems that need to use both types of data models.

Keywords: field-based model, object-based model, computability, managed objects, geographic space, 3d, time, scale.

1 Introduction

The geospatial branch of information science (GISc) has a dual approach to representing the real world. Its duality is manifested by 1) discrete objects, which are identifiable, countable entities existing in otherwise empty geographic space, and by 2) continuous fields, which for every location in a given domain of geographic space have a value forming a field of certain quantity. The twofold nature of the object / field modelling is a concrete, classical problem that has been eluding specialists in GISc since 1980 (Goodchild, 2012). This work studies the comparative suitability of object-based and field-based representations for computable geospatial models. The goal is to provide a theoretical and engineering solution supporting both field- and object-based geospatial models in a uniform way. This is the scope depicted in Figure 1. The problem in principle is that the solution must provide a better modelling flexibility without hindering simplicity and applicability in practice. Due to these unmet requirements the solution to duality of fields and objects might also address number of other practical issues, such as dealing with heterogeneity of models across many application domains, design and management of complex geospatial models, or difficulties with handling and exchange of the models by various information systems.

Computability is essential for this work since computable geospatial modelling is the main subject, but it is also important in a much broader sense because computable models can do work that humans would have to do otherwise. The rest of this section provides a necessary introduction to the computability theory. This gives a basis to Section 2 where the method for geospatial modelling is formalized and the engineering design presented. Section 3 reports on implementation related topics and on the resulting technology.

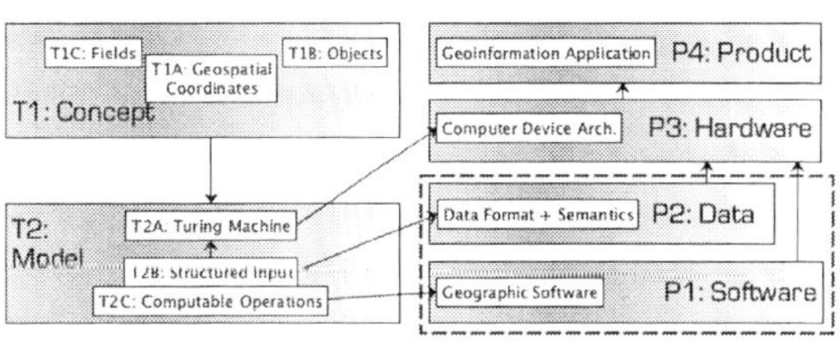

Figure 1: Computable geospatial modelling in context.

1.1 Computable models

This work deals with geospatial models that are *computable*. Computable in the sense that humans could also calculate the model manually. This means that the set of instructions followed to carry out a computation must be finite, that the computation carried is real - not imaginary (must be effective), and it must be possible to show how exactly the computation is performed (must be constructive). Exactly like in a program or a recipe.

The theory of computation is based on the concept of Turing machine. It implies, for example, that any finite set of Turing machines can be represented by a single one. Turing machines define a unique and natural class of "computable", which is fundamental for this work, but also for all computer science and mathematical logic.

Therefore, according to Sipser (1996) any computable model M can be formalized as Turing machine by a six-tuple:

$$M(\mathbf{Q}, \mathbf{\Sigma}, \delta, q_0, q_A, q_R) \tag{1}$$

where: $\mathbf{Q}$ is a finite set of all possible computational states of the model; $\mathbf{\Sigma}$ is the alphabet used to describe any possible instance of the model; δ is the *transition function* that tells how the model is evaluated between two consecutive steps; $q_0 \in \mathbf{Q}$ is the start state; $q_A \in \mathbf{Q}$ is the accept state; and $q_R \in \mathbf{Q}$ is the reject state.

The central element that makes the model "work" is the computable function δ also called as algorithm.

$$\delta : \mathbf{Q} \times \mathbf{\Sigma} \quad \rightarrow \quad \mathbf{Q} \times \mathbf{\Sigma} \times \{L, R\} \tag{2}$$

Given a *model instance* $\mathbf{I}$, which is an input string over alphabet $\mathbf{\Sigma}$, the computable function δ of M works as follows: for every valid state $q \in \mathbf{Q}$ and input character $c \in \mathbf{\Sigma} \wedge c \in \mathbf{I}$, the function returns the next state, writes (outputs) a character, and moves on reading the next character to the left L, or to the right R, from the current position in the input string $\mathbf{I}$. This continues forever unless δ yields the accept state q_A, or the reject state q_R.

Therefore the input is a model instance $\mathbf{I}$, which should not be confused with the model M itself. The M refers to a more general solution of the modeled phenomena. In contrast, the model instance $\mathbf{I}$ is a rather concrete utterance for which M computes an output.

For example, consider a model M_A deciding a "distance between two points on Cartesian plane". The model instance (input) is a pair of Cartesian coordinates, e.g. $[0, 0]$ and $[2, 0]$, and the decision (output) for this instance is 2, which is the distance between the points. A different model M_B would be necessary for "rendering the line between two points on a display". While the instances could be reused from the previous example, the transition function δ and output would be different in this case. Also, the model M_B may depend on yet another model, for instance representing the display. Such model hierarchy is common in practice and often gets complex.

The computable model can be understood as a compact collection of the problem-solving model M together with one or more distinct instances $\mathbf{I}$ solvable by M. Considering and treating computable model this way as a single, compact unit provides a key postulate in the modelling method suggested in Section 2. The content unit based on this view of M then has the ability to store not only the data $\mathbf{I}$ and the intermediate results of the computation, but also to store the transition function δ that brought about the computation. In Figure 1 is such compact unit depicted as *T2*, in which *T2B* would correspond to the model instance $\mathbf{I}$ and *T2C* to the transition function δ. The introduction of established computability theory formulated by Equation 1 is essential for addressing the concepts of space, fields and objects in the geospatial modelling method.

2 Method

A new method for geospatial modelling is based on the equivalence of finite composition of computable models M mentioned in Section 1.1. This composition equivalence can be expressed as:

$$M = (M_1, ..., M_n) \mid n \geq 1 \tag{3}$$
$$\delta = (\delta_1, ..., \delta_n) \tag{4}$$

Note that δ_1 is the transition function of M_1, δ_n is the transition function of M_n, and therefore δ is the function(s) of the computable model M. The principle of Equation 3 is applied to a universal geospatial model supported by the definition of geographic space.

2.1 Geographic Space

Every information system is constituted by a unique space and by contents associated with that space. This is true of any space. Not only abstract spaces used in database systems for querying and indexing of data, but any space with certain order, can be used as a basis for an information system. This includes the representation of the real-world space where the Earth exists. The conception of contents here are the computable models and their instances, but the space must be defined first. Space is critical for an information system because points of the space are the means of accessing the content. For example the World-Wide-Web (the Web) employs a non-topological, URL address space, which makes it possible to access a concrete Web page. Without the URL space the Web cannot exist.

The idea of devising a unique definition of space for all geospatial modelling is easy to express, but a good, general solution is missing. The definition of geographic space must sufficiently correspond to physics, but at the same time must facilitate the nature of the computable content. Modern geoinformation applications utilize many different spaces with two, three, and in rare cases with four and more dimensions resulting in more than thousand different coordinate systems used in practice (EPSG, 2013). The arguments for unique geocentric space are well known (Burkholder, 2000, Kjems & Kolar, 2005), but a definition that supports various geospatial models and that can be adopted by a broad variety of applications at different dimensions and scales still remains to be found.

The majority of geospatial applications can use the Newtonian description of space to obtain values that are correct to a sufficiently high order of accuracy. The realization of the geocentric reference frame, however, uses astrometry operating at the angular resolution exceeding one milliarcsecond, or atomic clock ticking at nanosecond level and moving on high precision orbits. Modelling events at such precision level, high speeds and gravitation differences requires consideration of general relativity (Kopeikin, 2007), and relativistic corrections must be used. Neglecting the relativistic curvature of space in these cases would degrade the observed measure (Pogge, 2013).

The *geographic space* is represented through a reference system, which relates coordinates of points existing in reality to a unique and common basis for computational geospatial models. The coordinates of the geographic space **S** have six-dimensions:

$$\mathbf{S} = \mathbf{X} \times \mathbf{Y} \times \mathbf{Z} \times \mathbf{T} \times \mathbf{\Gamma} \times \mathbf{\Omega} \quad (5)$$

The space **S** is defined close to the Earth and is co-rotating with it. Following the Newtonian physics the space is considered as Euclidean using Cartesian right-handed coordinates **C** with the same unit used for each dimension.

$$\mathbf{C} = \mathbf{X} \times \mathbf{Y} \times \mathbf{Z} \subset \mathbf{S} \quad (6)$$

The origin is close to the Earth's center of mass (geocenter), the orientation uses **X** and **Y** coordinates to represent the equatorial plane and the $z+$ axis is the direction of the north pole and of the Earth's rotation axis. In addition to Cartesian coordinates, other coordinates, e.g. geographical coordinates, could be used (Boucher, 2001). The temporal coordinates **T** are associated with the time running at the reference geopotential level (geoid) of the Earth.

It was mentioned that the coordinates are the means of accessing contents. However, due to the continuity of the Euclidean space, where all points are topologically relevant to each other, obtaining geospatial models for a given coordinates leads to a progressive loading of contents from the entire space (all the contents). Following the principle that closer things are more relevant than distant things the geographic space **S** has two *proximity dimensions*. The proximity dimensions in Equation 5 are *time proximity* $\mathbf{\Omega}$, and *spatial proximity* $\mathbf{\Gamma}$.

The spatial proximity $\mathbf{\Gamma}$ represents the extent of geometric neighborhood, and indicates a relative geometric magnitude within the space **S**. Given a position $p \in C$ the spatial proximity $\mathbf{\Gamma}$ allows to decide the range of relevant neighborhood for models of various geometric magnitude (size) on the scale from the largest to the smallest geometric magnitude. The *spatial proximity level* $\gamma = 0$ is for models of the largest geometric magnitude, $\gamma = 1$ is for smaller and so on. The levels of spatial proximity are obtained by a recursive octant subdivision (Weisstein, 2103) of the Euclidean space **C**. In order to perform computationally such octant subdivision bounds for the ranges of coordinates **X**, **Y** and **Z** must be specified. Choosing the same symmetric range for all three coordinates implies that the shape of the geometric space **C** is a cube, and its center coincides with the geocenter. This "root" cube also corresponds to the spatial proximity level $\gamma = 0$.

Analogously the time proximity $\mathbf{\Omega}$ is a dimension that facilitates decision about which models are relevant for a given instant on scale ranging from the shortest-term duration to the longest-term duration. The levels of time proximity are obtained from the interval of all time coordinates **T** by its binary recursive subdivision.

2.2 Geospatial Models

A computable geospatial model has a concrete formulation, but it is an abstract entity that can be applied to an arbitrary more concrete geospatial model. These models can be object-based, field-based, or complex including models combining both approaches. All computable geospatial models are defined in the geographic space **S** introduced in the previous section and carry several common properties. With consideration of Equation 3 the computable geospatial model G is expressed as a three-tuple:

$$G = (M_{sdx}, M_{ref}, M_{fun}) \quad (7)$$

The key part is the *functional model* M_{fun}, which can represent any computable model, again by following the concept behind Equation 3. While traditional solutions strictly separate the functionality δ from data $\mathbf{I}$ this method, in contrast, allows structures of data together with functionality. Together they form the content that is exchanged between applications. Although this perspective is common to programmers it is unfamiliar to content providers who store and exchange data. The traditional separation is depicted in Figure 1 by $P1$ and $P2$. Using this method means fusion of $P1$ and $P2$ into a single entity as denoted by the dashed line in Figure1. This is a significant concept and a key to the uniform approach to object-based and field-based geospatial models.

The second term in Equation 7 is the *reference model* M_{ref}, which provides values that reference the model G in the geographic space $\mathbf{S}$. There are three types of reference values:

$$P_{ref}(x, y, z) \in \mathbf{C} \mid P_{ref} \in M_{ref} \quad (8)$$

$$\gamma_{ref} \in \mathbf{\Gamma} \mid \gamma_{ref} \in M_{ref} \quad (9)$$

$$T_{ref}(t_{start}, t_{end}) \mid t_{start} < t_{end},\ T_{ref} \in M_{ref} \quad (10)$$

The reference point P_{ref} specifies Cartesian coordinates with which the model G is associated. The spatial proximity γ_{ref} indicates the metric size of both the model and its spatial neighborhood. Every model in the geographic space $\mathbf{S}$ exists for a limited period of time bounded by temporal coordinates $t_{start} \in \mathbf{T}$ and $t_{end} \in \mathbf{T}$. The value T_{ref} represents these *temporal bounds*. The only mission of M_{ref} is to return these reference values whenever needed.

2.3 Geospatial Index

According to Equation 7 each geospatial model G includes *the model of geospatial index* M_{sdx}. The geospatial index has two components with close relationship to the spatial proximity $\mathbf{\Gamma}$, and to the time proximity $\mathbf{\Omega}$ introduced in Section 2.1. They facilitate several properties common for all geospatial models. The spatial index c_{sdx} maps from P_{ref} and γ_{ref} to a set of *index coordinates* $\mathbf{C}_{sdx}$ by:

$$c_{sdx} : \mathbf{C} \times \mathbf{\Gamma} \rightarrow \mathbf{X}_{sdx} \times \mathbf{Y}_{sdx} \times \mathbf{Z}_{sdx} = \mathbf{C}_{sdx} \subseteq \mathbf{C} \mid c_{sdx} \in M_{sdx}. \quad (11)$$

Every $\mathbf{C}_{sdx}$ is a subspace of the Euclidean space $\mathbf{C}$ in such a way that it coincides with some octant generated by the subdivision of the spatial proximity $\mathbf{\Gamma}$. Given a spatial proximity level $\gamma = k \in \mathbf{\Gamma}$ there are 8^k distinct octants denoted as $\mathbf{C}^k_{sdx}$. All $\mathbf{C}^k_{sdx}$ have equal size, never intersect each other, and their sum fills the entire space $\mathbf{C}$. It is assumed that the reference point is one of the index coordinates $P_{ref} \in \mathbf{C}^k_{sdx}$. The signature of each octant $id(\mathbf{C}_{sdx})$ is used for a redundant indexing structure that facilitates rapid access to spatially relevant models and can also provide a paging mechanism for the access. Because $\mathbf{\Gamma}$ subdivision that generates $\mathbf{C}_{sdx}$ is space-driven rather then content-driven the indexing is applicable in distributed, decentralized information systems. When considered as local coordinate system $\mathbf{C}_{sdx}$ can be also utilized for more data-efficient geometry representation because fewer significant digits can be used. In this sense $\mathbf{C}_{sdx}$ also facilitates visual rendering and multiple levels of resolution because utilizing a fixed number of significant digits over large range and over smaller subsection leads to different resolution over these spatial domains.

The temporal index component is represented by function t_{sdx} that maps the temporal bounds T_{ref} to a unique interval of the linear time coordinates $\mathbf{T}_{sdx}$ as follows:

$$t_{sdx} : \mathbf{T} \times \mathbf{T} \ \rightarrow \ \mathbf{T}_{sdx} \subseteq \mathbf{T} \mid t_{sdx} \in M_{sdx}. \quad (12)$$

The signature $id(\mathbf{T}_{sdx})$ is utilized for fast access to temporally relevant models, which is an analogy to the use of $id(\mathbf{C}_{sdx})$ for indexing purposes. Because the Euclidean space $\mathbf{C}$ and the time $\mathbf{T}$ are independent dimensions of the geographic space $\mathbf{S}$ it is convenient to combine c_{sdx} and t_{sdx} in the single indexing model M_{sdx}.

2.3.1 Geospatial Object

An object-based representation of any geographic feature that is computable can be described using the geospatial model G expressed in Equation 7, and is called computable *geospatial object*. As used in the introduction, "object-based" is meant as a contrast to the field-based representation of geographic features, which will be addressed in the next section. It must be stressed that an abstract word "object", especially when computer science and engineering is involved, can easily lead to a confusion. The context to which the term "object" is related must be clarified. Here the geospatial object refers to an object related to the geographic space $\mathbf{S}$. We will see in Section 2.4 that the geospatial model itself can be called as object, but in context of an engineering solution to computable models - hence completely different type of object compared to the geospatial object.

The simplest geospatial object can be described by Equation 7 in which the functional model M_{fun}

does nothing. Such model of geospatial object provides the reference values P_{ref}, γ_{ref}, T_{ref}, and through the spatial index c_{sdx} and temporal index t_{sdx} they also have access to their proximity coordinates $\mathbf{C}_{sdx}$ and $\mathbf{T}_{sdx}$. If any of the reference values are missing then either geospatial model G cannot be made, or appropriate default values must be specified or generated by the application.

Any geospatial object can, however, through the functional model M_{fun} use other models arbitrarily, for instance dealing with various geometric types, topological operations, exchange data formats and other subjects as depicted in Figure 2. Note that the geospatial model is not only an entity in the system diagram, but also a content unit that can be stored and exchanged. The relatively simple representation described by Equation 7 is used, regardless of the specialization or complexity behind a particular geospatial object, . If an instance requires any specialized model that is part of M_{fun} it will be provided by the geospatial model G. That provides almost arbitrary flexibility to geospatial modelling while keeping the actual computable geospatial model G relatively simple for adoption by broad variety of applications. All the properties of geospatial objects described in this section also apply to the field-based models described in the following section.

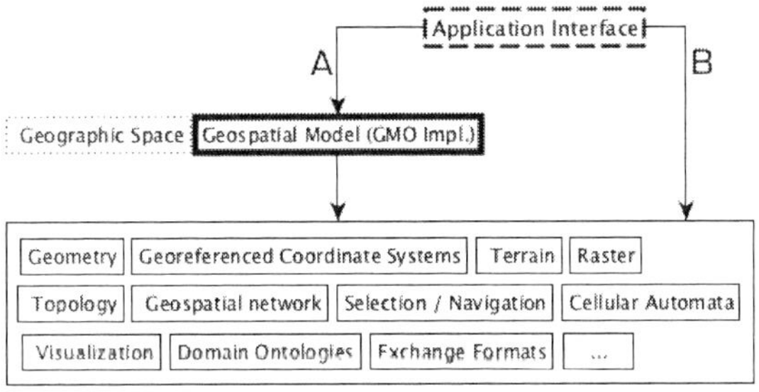

Figure 2: Geospatial model in relationship to the application interface.

2.3.2 Geospatial Field

A field-based representation of any computable geographic feature can be described using the geospatial model G from Equation 7, and is called computable *geospatial field*. The functional model M_{fun} of every geospatial field is characterized by a function f that maps from a spatial domain $\mathbf{D}$ to a field $\mathbf{V}$ as follows:

$$f : \mathbf{D} \rightarrow \mathbf{V} \times \mathbf{S} \mid f \in \delta_{fun} \wedge \delta_{fun} \in M_{fun} \quad (13)$$

The domain $\mathbf{D}$ is a geometric space, which may be a subset of the geographic space $\mathbf{S}$. The function f associates every point from $\mathbf{D}$ with value $v \in \mathbf{V} \times \mathbf{S}$, which means that the value v is also associated with the geographic space $\mathbf{S}$. The values $\mathbf{V}$ can be of any non-variable kind as long as it is computable by the function f. The function f must be expressed as part of the transition function δ_{fun} of the functional model M_{fun}. The extra (relatively to geospatial objects) function f for geospatial fields is required for the reason of computability and aggregates discrete values v over computable representation of space using discrete coordinates.

Every field-based model requires a certain domain $\mathbf{D}$ and a the type of field values. A good example of such domain is a three-dimensional Euclidean space with computable value v being the Cartesian coordinates $(x, y, z) \in \mathbf{C}$. Note that in this example both $\mathbf{D}$ and $\mathbf{C}$ are representations of three-dimensional Euclidean space, but their definition might be different as long as $\mathbf{D}$ allows function f to be computable. Hence the domain $\mathbf{D}$ might cover the geographic space $\mathbf{S}$ entirely or partially, but will always be a redundant representation of $\mathbf{S}$ specific to only some model G. The resulting field of coordinates would contain a set smaller or equal to all coordinates of the geographic space $\mathbf{S}$.

If we consider a geospatial model G with an empty M_{fun} as the simplest case, one can argue that the object-based approach is superior to the field-based approach. This is true in the context of this method. All "computable" is based on a finite set of discrete steps calculated by a discrete model (see Section 1.1), which effectively prevents making truly continuous representations. We always end up 1) with many discrete values within a geospatial model, or 2) with many discrete geospatial models. Either case manifesting the principles of object-based. At best fields can be represented using a set of computable coordinates that sample the field sufficiently. Quite similarly to the coordinates of the geographic space $\mathbf{S}$, but with an important difference that all the coordinates of the field must be evaluated, which in computable modelling always depends on some transition function δ.

Since model G allows for an arbitrary function δ under M_{fun} an arbitrary computable field-based model can be achieved. A true unification method for object-based and field-based modelling cannot a priori discount either approach. This requirement is fulfilled, but on the basis that geospatial model G with empty M_{fun} has object-based properties, and therefore each computable geospatial field inherits these object-based properties resulting in two layers of structure: first an object, then a field structure. Both approaches share the resolution limit of the geographic space coordinates **S**.

2.4 Engineering Hierarchy of Computable Models

The design and engineering aspects must be addressed In order to facilitate practical creation and exchange of the computable geospatial model G, which has been presented in theory. This section introduces four engineering layers considered for the realization of Equation 7 together with its conceptual design.

Each computable model is expressed as a separate Turing machine M - a theoretical computer. Manufacturing a special device for every computable model would make this method impossible to apply, but because of Equation 3 this requirement can be avoided. The design of general-purpose computers, first addressed in Burks et al. (1946), utilizes the equivalence of finite composition of computable models expressed in Equation 3. A general-purpose computer is a single electronic device that allows to load and run different computable models.

The assembly of electronic circuitry that can run computable models is the first engineering layer to consider. It is denoted as $L1$ in Figure 3. The central processing unit (CPU) chip $L1A$ runs *machine code*. The general-purpose computer uses machine code that keeps both the transitional functions δ (algorithms) and the model inputs **I** (data) of computable models together. A special kind of an in-chip model that can start loading computable models from an attached secondary memory is denoted as $L1B$. This loader $L1B$ is the first computable model that is loaded and run automatically on every computer $L1A$ start-up.

When successful, the loader $L1B$ passes the control over loading and starting of computable models to the next engineering layer depicted in Figure 3 as $L2$. Layer $L2$, which is called an operating system kernel, makes it easier to make and run programs by abstracting attached physical devices by computable models called drivers. Drivers together with the loader model denoted as $L2B$ are the main interfaces used by programs, and introduce differences between operating systems. The loader $L2B$ can manage the machine code included in programs and start them.

The third engineering layer $L3$ consists of programs that require a particular kernel for their execution. Note that "program" is yet another term for a computable model following Equation 3. Figure 3 shows three relevant types of programs; 1) geoinformation program $L3A$ that can utilize the computable geospatial model G (see Equation 7); 2) compiler $L3B$ that can translate source code into an executable code for certain operating system or virtual machine (VM); and 3) VM $L3C$ that implements similar features as entire layer $L2$ for the sake of making programs independent of a particular kernel. Hence VM $L3C$ also has a component $L3D$ that can load and start programs.

The fourth layer $L4$ depicted in Figure 3 contains programs that require a particular VM $L3C$ for their execution, but can (in principle) be independent of layer $L2$. Such independence is achieved by an intermediate code that is executed by the VM $L3C$. The independent executable code is called *bytecode* (dictionary.com, 2013) and is the key engineering concept used for the unification of geospatial computable models. The independence of bytecode from operating systems is attractive because it allows computable models to be portable over a broad variety of devices and systems in a similar manner as data in traditional exchange formats. Another advantage of VM based on bytecode is that the functionality δ can be coded using many input languages. The layer $L4$ conceptually mimics the layer $L3$, but for this work is necessary only the geoinformation program $L4A$ that can utilize, on the basis of bytecode, the computable geospatial model G from Equation 7.

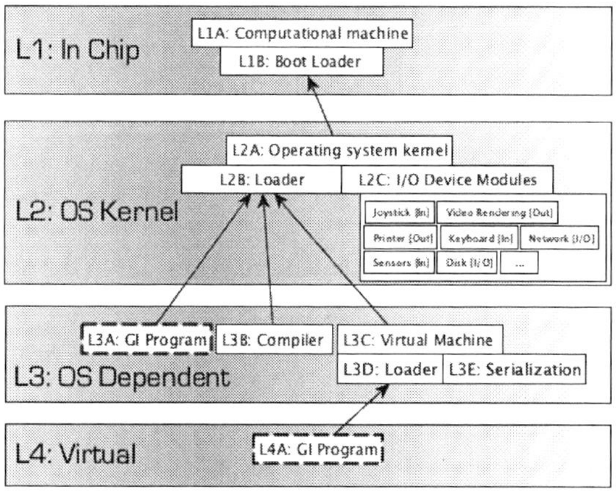

Figure 3: Engineering Hierarchy of Computational Models.

2.5 GMO: Geospatial Managed Object

The engineering hierarchy of general computable models suggests two nodes where the geospatial model G can be implemented. The two nodes are highlighted in Figure 3 as $L3A$ and $L4A$, and the dashed line denotes the interface to the geospatial model G depicted in Figure 2. Due to the comparative properties of layers $L3$ and $L4$ the most flexible design is to implement the geospatial model G under the node $L4A$ in Figure 3. The on-demand provision of components from the functional model M_{fun}, which is the key concept described in Section 2.2.2, then rely on the loader $L3D$. The ability of storing the geospatial model G depends on the serialization[32] model depicted in Figure 3 as $L3E$.

Design employing the loader model $L3D$ and the serialization model $L3E$ is considerably more advanced then traditional designs for exchange of geographic data. The traditional exchange data formats enforce fixed predefined structure and exchange of functionality is usually impossible or very limited, in contrast the geospatial model G in form of bytecode allows for variable structure and exchange of functionality. The nodes $L3D$ and $L3E$ from Figure 3 provide the engineering solution for Equation 3. The computable geospatial model G implemented under the node $L4A$ using the loader $L3D$ and the serialization model $L3E$ is called *geospatial managed object* (GMO). GMO is the engineering design associated with this method.

Design similar to GMO can be implemented under the node $L3A$ from Figure 3, but with limited applicability in practice. It is also possible to implement the geospatial model G under the node $L4A$ in a way that is dependent on layer $L2$, or limited in serialization of the geospatial model G. Such designs cannot be called geospatial managed objects. The GMO option is the most powerful engineering design in terms of sustainability, portability, applicability, and ability to reuse the geospatial model G.

The term "object" in GMO refers to an engineering paradigm called "object-oriented paradigm" used for programing of computable models and systems in general. In contrast the "geospatial object" is from the context of the geographic space $\mathbf{S}$. Keeping this in mind, the GMO is an engineering entity that corresponds to the computable geospatial model G. Hence GMO can be used for modelling of "geospatial fields" as well as of "geospatial objects" (see Section 2.2.3 and Section 2.2.2.) The term "managed" in GMO refers to the management of both data and functionality at the level of content, which is the key property of the presented method. The management of functionality is more complicated than management of only data. The bytecode and VM model $L3C$ solves this "management" of data and functionality on a common basis.

2.6 Model Incompleteness

The method of geospatial model G fully utilizes the computability theory for modelling geospatial phenomena. GMOs provide the optimal engineering design, but certain conceptual limits cannot be avoided. They come in three levels: computability, complexity and incompleteness. For example real or irrational numbers cannot be used in computational models - only their representations. An attempt to enumerate numbers such as π or $1/3$ would take eternity. The computable transition function δ must be always reducible to a sequence of basic arithmetic and logical operations that are physically implemented in the CPU $L1A$.

Complexity limits are given by the time and memory resources needed for computation. The precision of computable numbers used by GMOs is one aspect, because too precise representations of numbers (too much data) may not fit into the memory or take too long to evaluate. Also, many problems are inefficient when described by the transition function δ resulting in an algorithm with complex set of problem-solving operations. For example, a geospatial model deciding the shortest path through given number of cities n starting and ending in the same point requires handling of $(n-1)!$ distances (Applegate et al., 2007). A computer capable of billion operations a second (THz) would compute the solution for $n = 20$ cities in about 3.8 years, and for $n = 25$ cities would already need almost twenty million years. In the logistics and telecommunication industry, this example is an essential algorithm. But because it is so hard to compute, the practical solutions must use approximations and heuristic assumptions, leaving the exact solution unknown. This example is not exceptional, there are hundreds of such complex problems (wikipedia.org, 2013). Due to their inherent complexity, many solutions to well defined models are computationally intractable.

The limits of computability and complexity might be surprisingly constraining, but the reality is even worse. In 1931 Kurt Gödel derived a mathematical proof showing that a consistent set of axioms cannot be complete. There he also showed that proof of consistency cannot be derived from the given set of axioms alone (Hirzel, 2000). This has a profound im-

[32]In terms of computer science the process of transforming computable models from the form suitable for execution to a form suitable for storage and exchange is called "serialization".

plication on computational geospatial models, which are in essence axiomatic systems using arithmetic. The Gödel's theory informally says the best we can do about any geospatial model is to assume its consistency and at the same time hope that it sufficiently addresses the problem we want to model. We have to accept that all important aspects may NOT be addressed by the model, because there is always some true statement that is relevant but that cannot be covered by the model (Devlin, 1998).

Let's consider an example of a geospatial model from Goodchild (2012) that shifts a representation of a polygon in 2d by a given vector. This is a typical object-based model, which might be incomplete for field-based features. The shift by a given vector can lead to a result when polygon intersects other polygon, which for cadaster boundaries or contourlines is an invalid result. It highlights again that field-based and object-based considerations are ubiquitous in geospatial modelling, but the main point here is the nature of incompleteness. Regardless the limits of computability and complexity there is always more specialized case for which a geospatial model is incomplete. Hence, each computable geospatial model G and all its applications are primarily (and naturally) incomplete, and only then functional, field- or object-based, complex, secure, elegant and so on.

These limitations are valid for any computational model. They are included due to their importance in order to provide more consistent description of the method. The nature of incompleteness is often underestimated in geoinformation science and neglected by geospatial industry despite its implications on sustainability and efficiency of geospatial information systems. The geospatial model G provides an effective and flexible way how to deal with the nature of incompleteness.

3 Implementation

The design of geospatial managed objects introduced in Section 13 has been implemented as a software library named *geospatial reference interface for Internet networks* (GRIFIN). This experimental library additionally implements practical features not addressed in this article including mechanisms for storage and retrieval of GMOs, exchange of GMOs over network, automatic 2d interpretation of 3d geometries, support for visualization, and API for client applications to use GMOs. The experiment is maintained at `http://grifinor.net`, and the author's intention is to provide the library to everyone under the GNU General Public License.

In order for the GMO implementation to be applicable and robust few already existing technologies have been applied. The most crucial in this regard is the selection of the VM technology. Note that the utilization of VM here is not a mere software engineering convenience for implementation or porting to different operating systems. The nature of VM, as described in Section 13, is an inherent part of the GMO method. The role of the VM's bytecode relatively to GMOs is similar to the importance of HTML relatively to Web pages. The consequence of a future change to a different VM would have an analogy in changing HTML to, let's say, PDF format - making all the previous content obsolete and nonfunctional. Hence requirements on the VM technology are relatively strict and include: non-proprietary solution, strong focus on backward compatibility, production quality with commercial leadership, and widely established availability. Given these priorities the HotSpot VM, which is the original VM used within the Java ecosystem, stands out as nearly unchallenged choice, despite its currently marginal availability on mobile platforms.

The GMO concept has been coded in form of an abstract class, which implements the referencing mechanism M_{ref} and an abstract method $manage$ representing M_{fun}. This guarantees access to the referencing mechanism and custom functionality for all implementing subclasses and their GMO instances. The geospatial index M_{sdx} introduced in Section 13 has been implemented as a hash function mapping from the geospatial coordinates S to an array of 32 bytes. Examples at `http://grifinor.net/examples` also provide references to the source code. In order to facilitate prototyping and re-use of GMO models most of the examples utilize Scala language and GRIFIN Shell (GShell), which allows for an interactive use. GShell can be used to create, manage and consume the geospatial content on GRIFIN platform. It has all management, server, and remote access features available from a uniform environment and provides a way to exchange and execute code on GRIFIN's distributed network. This follows the original vision of a space for collaboration on model development, and not just a one-way publishing medium for static, predefined, and hard-to-change types of geospatial information.

Since 2006 several projects used GRIFIN, and many different GMO models were implemented as both object- and field-based representations of the real world environment. Figure 4 depicts results from InfraWorld project (Kolar, 2010) in three different software clients.

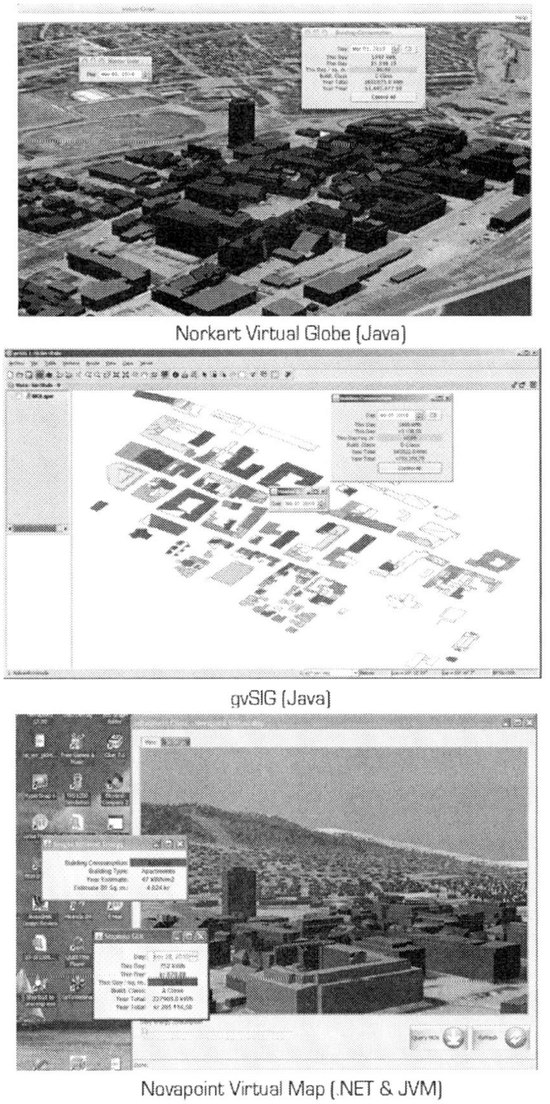

Figure 4: Three different software clients consuming identical GMOs from a server.

The city model and a daily energy consumption per building spanning a one-year-period were modelled as GMOs. While the model is relatively complex the software clients only implement the API for handling GMOs, which accounts circa twenty methods. The city model itself has several times more methods. The model brings all its functionality to the different clients, its specification and definition undertook big changes while the API for GMO could be implemented independently in parallel, and once the GMO API is done the software client is ready for all future models using the GMO method. This summarized three practical properties, which might be hard to address using non GMO solutions. Figure 5 depicts two examples of field-based representations. The right-hand side depicts a GMO model for "Nomenclature of Units for Territorial Statistics" including its hierarchical subdivision of administrative units. The spherical grids in Figure 5 show a GMO implementation of the Global Indexing Grid described in Kolar (2009), which is suitable for a spherical subdivision at arbitrary resolution and is a convenient basis for more complex models using the nearest neighbor queries. Energy City Frederikshavn (Wen et al., 2010) is other project based entirely on GMO method, featuring an object-based city model and a field-based terrain representation.

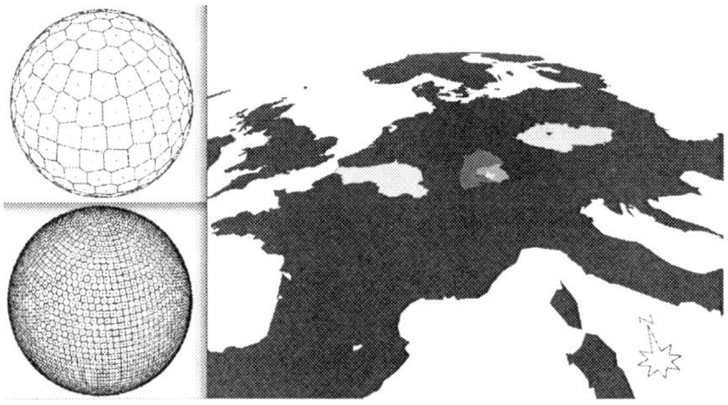

Figure 5: Field-based geospatial representations modelled using the GMO method.

4 Conclusion

A uniform theoretical and engineering solution supporting both field- and object-based computable geospatial models was presented and implemented. The geospatial model G and its GMO engineering design supports any computable representation of object-based or field-based geographic features, as well as complex geospatial models combining both approaches. The unification of the method leans on the definition of geographic space $\mathbf{S}$, which includes time, and which can be adopted at different scales and subdimensions. Scales are represented as unit-less dimensions of the space $\mathbf{S}$. Computable geospatial field inherits certain object-based properties enforced by the computability theory resulting in two layers of structure: first an object, then a field associated with the function δ. The key concept of the method is keeping the structured data together with the functionality δ as one content unit.

This is not new in general, but the method fully unfolds this potential of computability theory in GISc and brings it to geospatial applications. The bytecode, which is the engineering solution for keeping data and functionality together, is inherently associated with a particular VM. That is a unique requirement compared to solutions based on more traditional data exchange. GMO, as the optimal engineering design for the geospatial model G, encapsulates models on the level $L4$. This design is more flexible in addition and changes of models, more portable, and easier to deploy than solutions designed on the level $L3$ as well as those on $L4$ without the GMO encapsulation. GMOs make it possible to handle data and functionality at all ends of a distributed network, also after the object is instantiated, exchanged and consumed on the client side. This could facilitate the exchange of research developments of geospatial models. It was stressed that computational models are naturally incomplete, which has profound implications for the geospatial industry in terms of efficient sustainability of geospatial applications and for synergy of various information systems into larger infrastructures. Although the incompleteness is an unsolvable issue, GMOs provide an effective and flexible way to mitigate it. It also avoids many issues met by standardization efforts ongoing in the geospatial industry. The GMO design was implemented in the GRIFIN framework and is associated with the HotSpot VM. The VM requirement is the key engineering disadvantage of the method. Practice shows, however, that for the useful models the VM overhead can be minimized to a negligible level by applying adequate representations.

References

Applegate, D. L., Bixby, R. E., Chvatal, V. & Cook, W. J. (2007), *The traveling salesman problem: a computational study*, Princeton University Press.

Boucher, C. (2001), Terrestrial coordinate systems and frames, *in* 'Encyclopedia of Astronomy and Astrophysics', Nature Publishing Group, pp. 3289–3292.

Burkholder, E. F. (2000), The global spatial data model, *in* 'International Conference on Discrete Global Grids', Santa Barbara, California, USA.

Burks, A. W., Goldstine, H. H. & von Neumann, J. (1946), 'Preliminary discussion of the logical design of an electronic computing instrument', `http://grifinor.net/cite/2013/FOSS4G2013/4`.

Devlin, K. J. (1998), *The Language of Mathematics: Making the Invisible Visible*, W.H. Freeman & Company.

Dictionary.com (2013), 'Define byte-code'. URL: `http://grifinor.net/cite/2013/FOSS4G2013/2`

EPSG (2013), 'European Petroleum Survey Group: Geodesy Parameters Archived Versions'. URL: `http://www.epsg.org/archive.html`

Goodchild, M. F. (2012), 'Field-based spatial modeling', `http://grifinor.net/cite/2013/FOSS4G2013/3`.

Hirzel, M. (2000), 'On formally undecidable propositions of Principia Mathematica and related systems I', `http://grifinor.net/cite/2013/FOSS4G2013/7`.

Kjems, E. & Kolar, J. (2005), From mapping to virtual geography, *in* 'CUPUM', London, U.K.

Kolar, J. (2009), GIG: a projection-free global grid system suitable for indexing, *in* 'Proceedings of the 6th Symposium of the International Society for Digital Earth', Beijing, China.

Kolar, J. (2010), 'InfraWorld Project, The VERDIKT-Conference 2010', `http://blog.grifinor.net/post/3360221925`.

Kopeikin, S. M. (2007), Relativistic reference frames for astrometry and navigation in the solar system, *in* 'AIP Conference Proceedings', p. 268.

Pogge (2013), 'GPS and Relativity by Richard W. Pogge'. URL: `http://grifinor.net/cite/2013/FOSS4G2013/1`

Sipser, M. (1996), *Introduction to the Theory of Computation*, 1 edn, PWS Pub. Co., pp. 125–130.

Weisstein, E. W. (2013), '"Octant." From MathWorld–A Wolfram Web Resource'. URL: `http://gri[FB01?]nor.net/cite/2013/FOSS4G2013/6`

Wen, W., Kjems, E., Bodum, L. & Kolar, J. (2010), Dynamic features in a 3d city model as an energy system, *in* 'ISPRS Conference: International Conference on 3D Geoinformation', Berlin, Germany.

Wikipedia (2013), 'List of NP-complete problems'. URL: `http://grifinor.net/cite/2013/FOSS4G2013/5`